Creation

Hans Schwarz

William B. Eerdmans Publishing Company
Grand Rapids, Michigan / Cambridge, U.K.

© 2002 Wm. B. Eerdmans Publishing Co.

Wm. B. Eerdmans Publishing Co.
255 Jefferson Ave. S.E., Grand Rapids, Michigan 49503 /
P.O. Box 163, Cambridge CB3 9PU U.K.

Printed in the United States of America

07 06 05 04 03 02   7 6 5 4 3 2 1

Library of Congress Cataloging-in-Publication Data

Schwarz, Hans, 1939-
    Creation / Hans Schwarz.
       p.   cm.
    Includes bibliographical references and index.
    ISBN 0-8028-6066-4 (pbk.: alk. paper)
    1. Religion and science.   2. Creation.   I. Title.

    BL262.S38   2002
    231.7'652 — dc21

                         2002073888

www.eerdmans.com

Contents

Preface

The dialogue between theology and the natural sciences has gained momentum over the last few years. This is a good sign, because religion, if not related to the quantifiable, can easily deteriorate to superstition or ideology. The natural sciences on the other hand, when unrelated to religion, easily lose sight of the purpose of facts. An isolated fact is silent. It needs interpretation. Locating and grounding the fact in a particular context provides it with meaning.

While religion and the sciences are no longer on a collision course, still wanting is the mutual enrichment one might expect from meaningful dialogue between these worthy partners. Such mutual enrichment is largely wanting because we have separated God and the world to such an extent that God's interaction with the world is largely reduced to his interaction with our personal needs. For example, we affirm that God speaks to us and we pray to God, trusting that God is active. But then we conduct our lives in a world in which God does not seem to exist. We eagerly affirm that God did create the world. Yet we are usually at a loss to explain vis-à-vis our scientific knowledge what this actually means. Creation is usually reduced to God "pushing a button" at the beginning. Then the need for a creator vanishes. Creation therefore has little to do with the preservation of the world we inhabit. With regard to the world, God has been relegated to inactivity. But has theology really nothing to say about a God who is active in the world, in its natural course and in miraculous ways? Here the mere truce between science and religion needs to be turned into a mutually enriching experience. We must rediscover how the Creator is actively involved in the world in which we live. In our scientific endeavors we should become aware that we are not alone but that God is indeed the one who preserves the world and will lead it on its destined course.

The content of the subsequent pages has been presented many times in seminars and lecture courses both at the University of Regensburg and at Lutheran Theological Southern Seminary in Columbia, South Carolina. The material resulted in stimulating discussions. May these pages also stimulate the reader to consider how to relate our scientific discoveries to our faith in God the creator, sustainer, and redeemer. We need not retreat to a religious ghetto nor pontificate ideologically about our faith in order to give good reasons why we believe what we do believe.

At this point I would like to thank not just the students that helped to shape the content of these pages, but also Hildegard Ferme, who tirelessly and accurately prepared the various drafts, Dr. David C. Ratke for checking the footnotes, and Anna Madsen for proofreading and compiling the indices. Yet any shortcomings are still my own fault.

HANS SCHWARZ

Introduction

The Faustian drive,

> So that I may perceive whatever holds
> The world together in its inmost folds,
> See all its seeds, its working power,
> And cease work-threshing from this hour,

has become the leading motif of modern scientific knowledge.[1] Without ideological or religious premises one wants to understand the world as it is. Presupposed is that humanity is free and not subjected to any external strictures. At the same time Johann Wolfgang Goethe (1749-1832) wrote his poem *Faust*, the philosopher Immanuel Kant (1724-1804) demanded: "Have courage to use your own intelligence!"[2]

The real is only that which can be scientifically investigated. The results of scientific research can be proven; that is to say, under identical conditions one obtains the same results. The sciences, as conducted today, are descriptive and emphasize the facts. One does not ask about the essence of things, but attempts to depict processes or final states with the help of a formalized language. The world which is investigated is reduced to models so that instead of "substance" one talks about "func-

---

1. Johann Wolfgang von Goethe, *Faust*, part 1, ll. 30-33, trans. George Madison Priest, Great Books of the Western World (Chicago: Encyclopaedia Britannica, 1952), 11.
2. Immanuel Kant, "What Is Enlightenment?" in *The Philosophy of Kant: Immanuel Kant's Moral and Political Writings*, edited with an introduction by Carl J. Friedrich (New York: Random House, Modern Library, 1949), 132.

tion," and instead of "quality," "quantity." Whether it be psychology, biology, or physics, one attempts to comprehend and depict reality with the help of mathematical functions. This mathematization of the sciences, which is also becoming increasingly popular in the humanities, assures a certain degree of certainty which could not be obtained without mathematics. Already in 1921 Albert Einstein pointed to the problematic nature of such a procedure when he asked: "How can it be that mathematics, being after all a product of human thought which is independent of experience, is so admirably appropriate to the objects of reality? Is human reason, then, without experience, merely by taking thought, able to fathom the properties of real things? In my opinion the answer to this question is, briefly, this: — As far as the laws of mathematics refer to reality, they are not certain; and as far as they are certain, they do not refer to reality."[3]

If the sciences only respect as real that which can be comprehended by experimentation and examination, then they dangerously reduce reality, because that which is expressed by science is not identical with the whole of reality.

Up until the present many people identified reality with the results of scientific investigation. Consequently metaphysical issues were excluded, and if they still surfaced, they were synonymous with the quest for the objective. God either disappeared from nature or was equated with it. In the latter sense God was no longer prior to humanity and could still challenge humans, but since he was identical with that which they perceived, they could use the object of investigation in their own way. Humanity became the measure of all things and objectified nature could be used according to human reasonableness.

As we recognize more and more how dubious human knowledge has become, the question reemerges whether science as a human enterprise (which gives us validated knowledge) does not already interpret this knowledge in an illegitimately truncated way. This means that theology again has the possibility of entering the scene with its own understanding of reality and introducing this in the dialogue with the various sciences in a supplementary and sometimes even critical way. When we remember the officially materialistic worldview of communism, how-

---

3. Albert Einstein, "Geometry and Experience" (1921), in *Sidelights on Relativity* (New York: Dover, 1983), 28.

ever, we notice that until recently theology and the sciences were on a collision course. Nature was interpreted without any reference to God, or at least with the tacit or expressed confession that nature no longer needs a "God hypothesis."

Part I
Nature without God

The Dawning of Modern Science

Johannes Kepler (1571-1630), who discovered the two laws of planetary motion named after him; Galileo Galilei (1564-1642), who came into conflict with the Inquisition by advocating the Copernican heliocentric worldview; and René Descartes (1595-1650), who introduced radical doubt into philosophy, may serve as important representatives for the beginning of modernity. Though the beginning of modernity is often equated with the Copernican change from a geocentric to a heliocentric worldview, Nicolaus Copernicus (1473-1543), a canon of the cathedral chapter at Frauenburg, Poland, was more interested in developing a system which would reinstate the classical concept of harmony instead of breaking ground for a new view of the world. "He clearly had no intentions of abstracting his geometry from the actual motions of the heavens as such."[1] Copernicus even sacrificed accuracy for the sake of desired elegance, and thus it was not at all unreasonable that Galilei was told to teach the heliocentric theory as a hypothesis only and not as fact. It was only Kepler, following his mathematics and revolutionist thoughts of astronomy, who allowed his sense of harmony to be reformed by observation.

According to the traditional notion of harmony which Copernicus still cherished, Kepler's planetary orbits with two foci were considered rather "monstrous." Eventually, however, the heavens that "declare the

---

1. So Harold P. Nebelsick, *Circles of God: Theology and Science from the Greeks to Copernicus* (Edinburgh: Scottish Academic Press, 1985), 237.

glory of God" (Ps. 19:1) were understood to declare it in terms of creation and not of divine perfection. This implied that creation was not of divine material but of earthly reality with a contingent, rational order of its own. With this change from Greek harmony to Judeo-Christian matter-of-factness the mood was set for regarding the material world simply as creation. Since there were no longer divine qualities discernible in nature, the suspicion could now be nourished by believers and unbelievers alike that this could lead to atheism.

The Reformation, with its preoccupation with God's Word addressed to the individual, was not an opportune time for the development of a theology of creation. While both Martin Luther (1483-1546) and John Calvin (1509-64) emphasized the sovereignty of God, they did so to declare God not as creator but rather as the Lord of history. This meant that science could continue its course relatively unhampered by the salvational concerns of theology. Science did not seek out theology as a dialogue partner but pursued its task of describing nature's contingent, rational order, which it began to discover in ever greater detail. Once scientists discussed their findings in public, theology was taken by surprise and attempted to combat these "outrageous" theories. Galilei, Giordano Bruno (1548-1600), and Kepler got in trouble with their respective churches when they, particularly Galilei and Kepler, asserted the supremacy of science in matters concerning nature, and also in Bruno's case when he hinted that God did not work alone. *De revolutionibus orbium coelestium libri sex (Six Books on the Revolutions of the Celestial Bodies)* by the cautious Copernicus included a preface by the Lutheran theologian Andreas Osiander (1498-1552) in which he stated the strictly mathematical intention of this treatise.

A bifurcation had already begun that conceived of God as no longer being situated within the natural realm but beyond and outside it. Kepler, for instance, was advised: "Do not trust too much in your reason and see to it that your faith is founded on the power of God and not on human wisdom."[2] The Lutheran theologian Johann Arndt (1555-1621) was a refreshing exception. In his *Vier Bücher vom wahren Christentum (Four Books on True Christianity)* he showed in the fourth book, "The Book on Nature," that there is a correspondence between world and humanity,

2. "Konsistorium in Stuttgart an Kepler in Linz," in Johannes Kepler, *Gesammelte Werke* (Munich, 1955), 17:32.

4

macrocosm and microcosm, since nature is pure spirit and a symbol of God's activity.

It was not until the late seventeenth century that an interest in nature was reawakened in theologians, this time in the trappings of physico-theology. By then, however, the mathematization of nature had made significant progress. Descartes, for instance, aimed for scientific explanations which would be entirely mechanical with final causation totally excluded so that mathematical physics would emerge as the fundamental science. Yet he still had problems with a total mechanization of nature. This became evident when he confessed: "To demand of me a geometrical proof in a matter which depends on physics is to want me to perform an impossible task."[3]

In his *Philosophiae Naturalis Principia Mathematica,* begun in 1684 and published in 1687, Isaac Newton (1642-1727) used for the first time a single mathematical law to account for the phenomena of the heavens, the tides, and the motion of objects on earth. His mechanics guided astronomers and scientists in their search for natural knowledge. For Newton the mathematization of nature still revealed the glory of God. Thus he concluded his *Principia* with the assertion that "If the fixed stars are the centers of other like systems, these, being formed by the likewise counsel, must be all subject to the dominion of One. . . . This Being governs all things, not as the soul of the world, but as Lord over all. . . . It is allowed by all that the Supreme God exists necessarily; and by the same necessity he exists *always* and *everywhere*."[4]

Newton's conclusion, however, left many untouched: "To the eighteenth and much of the nineteenth centuries, Newton himself became idealized as the perfect scientist: cool, objective, and never going beyond what the facts warrant to speculative hypotheses. The Principia became the model of scientific knowledge, a synthesis expressing the Enlightenment conception of the universe as a rationally ordered machine governed by simple mathematical laws."[5]

Even the fundamental principles from which the system of the world was deduced seemed to be for some, such as Kant, an a priori

3. René Descartes, "Descartes à Mersenne, 17. Mai 1638" (157), in Descartes, *Correspondence,* introduction by Ch. Adam and G. Milhaud (Paris: Felix Alcan, 1939), 2:266.

4. Isaac Newton, *Mathematical Principles,* trans. Florian Cajori (New York: Greenwood, 1969), 2:544f.

5. Dudley Shapere, "Isaac Newton," in *Encyclopedia of Philosophy,* 5:491.

truth, attainable by reason alone. Initially, however, an optimism prevailed both in science and in relating scientific discoveries to the Christian faith. This mood seemed to be analogous to the optimistic reception of nineteenth-century Darwinian ideas in North America. One would have thought that the scientific penetration of nature would even more reveal the glory of God. For instance, the Dutch scholar, theologian, and administrator Bernhard Nieuwentyt (1654-1718) wrote in his monumental work *Het regt gebruik der werelt beschouwingen* (*The Right Use of the Understanding of the World*, 1715): "From all of this it follows that an accurate perception of that which we encounter in the physical world is a sure means of escaping the manifold causes of and opportunities for atheism and of perceiving the perfection of God in his works."[6] Similarly, Charles Bonnet (1720-93) concluded in his voluminous *Contemplations de la nature* (*Contemplations of Nature*, 1764-65): "This creator worthy of adoration must be unceasingly sought in the unfathomable chain of the various works of nature in which his power and wisdom is reflected with so much truth and splendor. He does not reveal himself to us immediately, since his plan which he has conducted would not allow such. But he has commanded heaven and earth to proclaim to us who he is. He has arranged our insights according to this divine language and raised sublime souls who search out such beauties and explain them."[7]

Yet all the different ichthyo- (fish), insecto-, litho- (stone), pyro- (fire), and astro-theologies which were developed in the wake of scientific discoveries did not do away with the fact that gradually God became less and less necessary for a nature which largely was conceived as an arrangement of geometrical shapes and numbers. While Newton still needed God to keep the stars from collapsing into one mass under the influence of gravitation and to maintain the stability of the solar system in the face of planetary perturbations, the need for such a "God of the gaps" was ever more reduced.

When the French mathematician and astronomer Pierre Laplace (1749-1827) had finished his monumental five-volume work *Mécanique céleste* (*Celestial Mechanics*, 1799-1825), he summarized the issue in his

6. Bernhard Nieuwentyt, *Het regt gebruik der werelt beschouwingen*, in *Klassiker des Protestantismus*, vol. 7, *Das Zeitalter der Aufklärung*, ed. Wolfgang Philipp (Bremen, 1963), 74f.

7. Charles Bonnet, *Betrachtungen über die Natur*, 4th ed. (Leipzig: Johann Friedrich Junius, 1783), 2:533.

famous reply to Napoleon's inquiry as to where the proper place for God was in his system: "Sir, I do not need this hypothesis." God was no longer necessary within a scientific worldview. The world made sense without any reference to God. Not even the hypothesis of the creator seemed necessary any longer. In 1842 the German physicist Julius Robert von Mayer (1814-78) formulated the first law of thermodynamics (or the law of conservation of energy), which states that within an energetically isolated system the amount of energy neither increases nor decreases. This law made it possible to endow the world with the attribute of eternity. Provided that the world is an energetically isolated system, it has no beginning and no end. It is eternal. Thus the starting point of a first creation and the God hypothesis of a first creator obviously become obsolete.

Once Charles Darwin (1809-82) had written his two epoch-making books *On the Origin of Species* (1859) and *The Descent of Man* (1871), the origin of humanity could be explained as part of the continuous evolutionary process within our world. There was nothing peculiar to humanity or to its ideas; they were only products of the evolutionary process out of which they originated. Thus a completely homogeneous worldview in atheistic terms seemed unavoidable. The destiny of religion in general and of the Christian God in particular seemed to be decided. The German zoologist and enthusiastic follower of Darwin, Ernst Haeckel (1834-1919), summed up this sentiment in his book *The Riddle of the Universe* (1899; ET 1900). With an uncompromising monistic attitude he asserted the essential unity of organic and inorganic nature. Just as the highest animals have evolved from the simplest forms of life, so the highest human faculties have evolved from the "soul" of animals. Such cherished ideas as the immortality of the soul, the freedom of the will, and the existence of a personal God were discarded. Haeckel suggested that those who still want to believe in God actually believe in a *"gaseous vertebrate,"* gaseous because God is worshiped as a "pure spirit" but not without a body, and a vertebrate because of our anthropomorphic conception of God.[8] In other words, Haeckel tells us that God is an impossibility, a contradiction in itself.

---

8. Ernst Haeckel, *The Riddle of the Universe at the Close of the Nineteenth Century*, trans. Joseph McCabe (New York: Harper, 1900), 288.

The Attack of Materialism

Approximately 150 years before Haeckel there emerged in France an atheism based on scientific and philosophical principles which soon had supporters in Germany and England and initially picked up on the atomism of antiquity. Some of the pre-Socratic philosophers, such as Democritus (ca. 460–ca. 370 B.C.), based their teaching regarding the world on the idea that there are atoms, meaning ultimate and indivisible building blocks of everything that is. According to Democritus, there is one uniform entity without any qualitative differences. It consists of smallest parts which are no longer divisible and therefore are called atoms. He explains the different qualities of things we feel as subjective impressions. They have no objective reality but only appear to us. Nature consists of atoms "which are thrown around in empty space."[9] Even the soul consists of atoms, and our thinking is a movement of atoms. Epicurus of Samos (341-270 B.C.) and his school renewed the atomism of Democritus. They advanced the idea that there is an infinite number of ultimate elements, which were no longer divisible and solid, called atoms. They have no quality and are only distinguished quantitatively through form and weight. Moreover, there is an empty space in which these atoms move. With these two elements, bodies and space, Epicurus attempted to explain all being. Even soul and spirit are called bodies, but bodies which consisted of finest matter. While soul and spirit are divisible and therefore as mortal as the body, the atoms are from eternity and will remain in eternity.

This kind of materialism was further developed in modernity especially by the French physician and philosopher Julien Offray de la Mettrie (1709-51). In his book *Man as Machine* (*L'Homme machine*, 1748), he presented a naturalistic view of humanity and explained spiritual processes through physiological causes. The soul, for instance, originates from the organization of the body, and the higher development of the reasonable human soul is due to the larger and more intricate development of the brain. According to la Mettrie, this thoroughgoing naturalism necessarily leads to atheism. Already in 1745 in his *History of the Soul* (*Histoire naturelle de l'âme*), he rejected metaphysical dualism and ex-

---

9. Democritus, *Fragment* 168, in *Fragmente der Vorsokratiker,* Greek and German, ed. Hermann Diels, 12th ed. (Dublin: Weidmann, 1966), 2:178.

plained the spiritual faculties through a motorlike power which resides in matter. The German baron Paul Heinrich Dietrich von Holbach (1723-89) offered similar explanations in his book *System of Nature* (*Système de la nature*, 1770). He described humanity as a product of nature which is subjected to the laws of the physical universe. Beyond that there are no further ultimate principles or powers. According to von Holbach, it is an illusion to consider the soul as a spiritual substance. The moral and intellectual attributes of humanity can best be explained in a mechanistic way through physical, biological, and social interactions. The empirical and rational exploration of matter provides for von Holbach the only possibility of understanding what a human being is all about. Nature is the sum of all matter and of its movement. Matter is actually — or at least potentially — in movement, since energy or power is a property innate in matter. The material universe is simply there. We need not pose the question concerning the creation of matter. There is neither accident nor disorder in nature since everything occurs out of necessity and in an order which is determined through the irreversible chain of cause and effect. The world in which we live is therefore not only interpreted in a mechanistic way, but von Holbach also believed it was subjected to a stringent causal determinism. This was the general mood at the beginning of the nineteenth century. Therefore it was no accident that in November 1793, in the wake of the French Revolution, God was officially abolished and in God's place the goddess of reason was enthroned.

Taking this brief summary into account, we are not surprised that the German philosopher Ludwig Feuerbach (1804-72) proceeded along the same lines in *The Essence of Christianity* (1841). He claimed to be "objective" and would use only "the method of *analytic* chemistry."[10] For Feuerbach theology could no longer be understood ontologically or mythologically; he regarded it "as psychic *pathology*." Theology, religion, and God are the result of a sick psyche, because humanity creates an image of itself which it projects into another world and in which it believes and on which it reflects. God corresponds to wishful thinking by which humanity hypostatizes its own ideals and adores it. Feuerbach therefore concluded: "God was my first thought, reason my second, humanity my

---

10. Ludwig Feuerbach, *Das Wesen des Christentums,* in *Gesammelte Werke,* ed. Werner Schuffenhauer (Berlin: Akademie, 1973), 5:6, including the following quote. Unfortunately this part of the introduction was omitted in the English translation.

third and last. The subject of the Godhead is reason, but the subject of reason is humanity."[11] Humanity produces reason and reason in turn produces God. Therefore everything begins and ends with humanity and that which can be analyzed and scientifically interpreted. Feuerbach did not want to understand religion and humanity in a vulgar and materialistic way, since he did not simply reject religion. But his idea of religion as "a human projection" decisively paved the way for Marx and Engels.

Friedrich Engels (1820-95) wrote: "All religion, however, is nothing but the fantastic reflection in men's minds of those external forces which control their daily life, a reflection in which the terrestrial forces assume the form of supernatural forces."[12] Religion is for him a projection of the external world in which we live onto a believed-in super-world. Humanity is the originator of religion; it is religion that shapes humanity. Since humanity can project its wishful dreams onto a believed-in heaven or a hereafter, it is unable to change the conditions which make such projections necessary. Therefore both Karl Marx (1818-83) and Feuerbach and the later Marxists and communists who followed in their wake vehemently criticized every religion and all beliefs in God. Marx claimed:

> The basis of irreligious criticism is: *Man makes religion,* religion does not make man. In other words, religion is the self-consciousness and self-feeling of man who has either not yet found himself or has already lost himself again. . . . *Religious* distress is at the same time the *expression* of real distress — and the *protest* against real distress. Religion is the sigh of the oppressed creature, the heart of a heartless world, just as it is the spirit of a spiritless situation. It is the *opium* of the people. The abolition of religion as the *illusory* happiness of the people is required for their *real* happiness. The demand to give up the illusions about its condition is the *demand to give up a condition which needs illusions.* The criticism of religion is therefore *in embryo the criticism of the vale of woe,* the *halo* of which is religion.[13]

---

11. Ludwig Feuerbach, "Philosophische Fragmente" (1843-44), in *Sämtliche Werke,* ed. W. Bolin and F. Jodl (Stuttgart: Frommann, 1959), 2:388.

12. Friedrich Engels, "Anti-Dühring" (Extracts), in Karl Marx and Friedrich Engels, *On Religion,* introduction by Reinhold Niebuhr (New York: Schocken, 1964), 147.

13. Karl Marx, "Contribution to the Critique of Hegel's Philosophy of Right," in Marx and Engels, *On Religion,* 41f.

Humanity must reflect on itself and on the conditions in which it lives, i.e., its environment. Any idealism is rejected, because "matter is not a product of mind, but mind itself is merely the highest product of matter."[14]

Marx and Engels were not alone in their materialistic and naturalistic attitude. They represented the main stream of the common thinking of that time. Indicative for this was the so-called materialism controversy of 1854-55, which was a decisive event in the intellectual and religious history of Germany. It started in 1852 with a literary exchange in the *Augsburger Allgemeinen Zeitung* between the Göttingen professor of medicine Rudolf Wagner (1805-64) and the materialistic physiologist Carl Vogt (1817-95). The latter had rejected in earlier publications the biblical tenet that humanity originated from one single human couple. In 1854, at the Thirty-first Convention of Natural Scientists in Göttingen, this controversy reached its climax when the ecclesiastically conservative Wagner postulated the descent of all humans from one single couple which at the same time resembled the ideal portrayed by the Indo-European race. Though he thought spiritual impressions and activities were correlated to the brain and the nerves, he attempted to show that this does not preclude the existence of a special substance called "soul" which cannot be weighed and is invisible. Therefore the existence of individual immortal souls cannot be excluded. His lecture, entitled "Concerning the Creation of Humanity and the Substance of the Soul," was followed in the same year by another publication, "Knowledge and Faith with Special Relationship to the Future of Souls." Vogt responded to these papers in 1854 with a publication entitled *Köhlerglaube und Wissenschaft (Backwoods Faith and Science),* which has been reprinted many times since. He rejected the thesis of a creator who interferes with the course of nature, and against the claim of an independent existence of the soul, he postulated a strictly physiological understanding of the soul. He contended that "all faculties which we subsume under the name of activities of the soul are merely functions of the brain or, to say it here in a rather drastic way: thoughts are related in the same way to the brain as bile is to the liver or urine to the kidneys."[15] All activities of consciousness must be traced back to activities in the

---

14. Friedrich Engels, "Ludwig Feuerbach and the End of the Classical German Philosophy," in Marx and Engels, *On Religion,* 231.

15. Carl Vogt, *Köhlerglaube und Wissenschaft* (Gießen: Rieker, 1855), 32.

brain, and a soul which is independent over against the body is forever rejected. Furthermore, he claimed that religious assertions cannot be proven by science. They are purely private opinions. By now materialism had won the upper hand in the mind of the majority, and Feuerbach could count on widespread agreement when he stated: "Humanity is what it eats."[16]

The Dutch physiologist and philosopher Jacob Moleschott (1822-93) had argued similarly against the more conservative Justus von Liebig (1803-73): "Without phosphorus there is no thought."[17] Similar to the physician Ludwig Büchner (1824-99), Moleschott claimed that energy and matter cannot be separated. There is no separate material foundation on which a spiritual world can be built, but matter can neither occur nor be conceived of without energy, neither can energy be conceived without matter. Yet Moleschott did not want to consider the material as dead matter, since there is always already an energy present in it, since life permeates the whole world and with it thought as well. In the nineteenth century materialism was by no means understood statically so that everything could be reduced to an unchangeable material foundation. To the contrary, like the political, societal, and industrial advances of the time, materialism had a dynamic structure. This materialism which Engels called "modern" perceives "in history the developmental process of humanity whose laws of advancement one has to discover."[18] Though this materialism rejects any kind of teleology which has its origin in God, a goal-directed movement was not totally excluded, since one had realized that humanity and the entire history of the world have continuously developed and are subject to a progressive evolution. Yet the idea of evolution, which had made its imprint on the nineteenth century long before Charles Darwin, contained — at least in Europe — an explicit attack on the Christian faith.

---

16. Ludwig Feuerbach, *Das Geheimnis des Opfers oder Der Mensch ist, was er ißt*, in *Gesammelte Werke*, 11:26.

17. Jacob Moleschott, *Der Kreislauf des Lebens*, 5th ed. (Giessen: Emil Roth, 1887), 2:227.

18. Friedrich Engels, *Herrn Eugen Dühring's Umwälzung der Wissenschaft*, in Karl Marx and Friedrich Engels, *Werke* (Berlin: Dietz, 1968), 20:24.

The Attack of Evolution

The term "evolution" is indelibly connected with the name and person of Charles Darwin. Yet the idea of a development of the living species and of the world they inhabit had been widely accepted long before Darwin. He neither used the term "evolution" nor was the first to propound evolutionary ideas. Most of the ideas associated with him, such as "the struggle for survival," "selection according to sex," and "the survival of the fittest," he took over from others. His grandfather, Erasmus Darwin (1731-1802), a renowned poet and physician, had already suggested that "perhaps millions of ages before the commencement of the history of mankind [...] all warm-blooded animals have arisen from one living filament, which *the great First Cause* endued with animality, with the power of acquiring new parts, attended with new propensities, directed by irritations, sensations, volitions, and associations; and thus possessing the faculty of continuing to improve by its own inherent activity, and of delivering down these improvements by generations to its posterity."[19]

In 1809 the French zoologist Jean-Baptiste de Lamarck (1744-1829) published his *Zoological Philosophy*. In it he also provided a clearly evolutionistic view of life. However, he attempted to explain evolution by pointing to the cumulative inheritance of modifications induced by environmental influence. These modifications can become constant and lasting and even lead to the modification of old organs or to the need for new ones. The primary and best-known example for Lamarck's point is the giraffe. Lamarck claimed that since the giraffe lives in places where the soil is nearly always arid and barren, "it is obliged to browse on the leaves of trees and to make constant efforts to reach them. From this habit long maintained in all its race, it has resulted that the animal's forelegs have become longer than its hind legs, and that its neck is lengthened to such a degree that the giraffe, without standing up on its hind legs, attains a height of six meters."[20]

---

19. Erasmus Darwin, *Zoonomia or the Laws of Organic Life* (39.4.8), 3rd ed. (London: J. Johnson, 1801), 240. It is astounding that Charles Darwin never mentioned his grandfather even though the latter postulated many theses more precisely than his famous descendant.

20. Jean-Baptiste de Lamarck, *Zoological Philosophy: An Exposition with Regard to the Natural History of Animals*, trans. Hugh Elliot (Chicago: University of Chicago Press, 1984), 122.

13

This idea that environmental influences significantly affect the evolutionary process was later picked up by Marx and Engels, who believed that this type of inheritance could be used to facilitate future improvements of the human race.[21]

Another important precursor of Darwin who must be mentioned is the British economist Thomas Robert Malthus (1766-1834). In his famous *Essay on Population*, the first edition of which appeared in 1798, he suggested that the human race always tends to outrun its means of subsistence and thus can only be kept in bounds by famine, pestilence, or war, or through prudential checks, such as postponement of marriage.[22] Charles Darwin confesses the strong impact Malthus had on his thinking when he writes: "In October 1838 . . . I happened to read for amusement Malthus on *Population*, and being well prepared to appreciate the struggle for existence which everywhere goes on from long-continued observation of the habits of animals and plants, it at once struck me that under these circumstances favorable variations would tend to be preserved, and unfavorable ones to be destroyed. The result of this would be the formation of new species. Here, then, I had at last got a theory by which to work."[23]

Yet Darwin was not a plagiarist. He wanted to make sure that the ideas he received from Malthus could be substantiated by observable data. He conducted experiments, read, and traveled. Finally by 1844 "Darwin had convinced himself that species are not immutable and that

21. For details see Friedrich Engels, *Dialektik der Natur: Notizen und Fragmente*, in Marx and Engels, *Werke*, 20:564, where he claims that Darwin lumped together two completely separate ideas, selection under the pressure of overpopulation and selection through the greater capacity to adapt to altered circumstances. A bizarre and unfortunate example of the exclusive emphasis on environmental influence was the rule of Lyssenkoism in the former Soviet Union. Trofim Denissowitsch Lyssenko (1898-1976) advanced a theory concerning the inheritance of acquired traits which later proved to be wrong. According to Lyssenko, genetic dispositions are a concentration of environmental conditions which in plant organisms have been acquired in a series of preceding generations. This would mean that a change of the environment would also bring with it a change in the species. If this theory is then extended to humanity, humanity would be indeed, as Marxism had claimed, largely a product of its societal conditions.

22. Cf. Thomas Robert Malthus, *Population: The First Essay*, introduction by Kenneth E. Boulding (Ann Arbor: University of Michigan Press, 1959), 21ff.

23. Charles Darwin, *Autobiography*, ed. Nora Barlow (New York: Harcourt, Brace, 1959), 120.

the main cause of their origin was natural selection, but he continued to work year after year to gain yet surer evidence."[24] His famous book *On the Origin of Species by Means of Natural Selection, or the Preservation of Favored Races in the Struggle for Life* may well have never been published if there had not been outside pressure.

The English naturalist Alfred Russel Wallace (1823-1913) had also come across Malthus's *Essay on Population,* and this book triggered in him the idea of the "survival of the fittest." Inspired by Malthus, Wallace sent a paper to Darwin, who recognized that it had striking similarity with his own ideas. Yet Darwin was fair enough not to claim priority over the ideas of Wallace, who was as yet completely unknown to Darwin. Upon the advice of his friends, he sent Wallace's paper to the Linnean Society (named after the Swedish physician and scientist Carl von Linné [1707-78]) together with an explanatory letter to the secretary and an abstract of his own theory written in 1844. Both papers were read in 1858 at the Linnean Society and published in its journal. Now Darwin saw the time ripe for his own ideas, and as the result of more than twenty years of work he published *The Origin of Species* in 1859. The main tenets expressed in his book are:

1. There are random variations among species.
2. Populations increase at a geometrical rate and, as a result, there is a severe struggle for life at one time or another.
3. Since there are variations useful to organic beings, individuals with useful variations "will have the best chance of being preserved in the struggle for life."
4. Individuals with useful variations will pass on the beneficial traits to the next generation and "will tend to produce offspring similarly characterized."[25]

Thus Darwin arrived at the idea of the survival of the fittest by means of natural selection. Though he claimed that this leads to an improvement of each creature in relation to its organic and inorganic con-

24. Sir William C. Dampier, *A History of Science and Its Relations to Philosophy and Religion,* postscript by I. Bernard Cohen, 4th ed. (Cambridge: University Press, 1966), 277.

25. Charles Darwin, *The Origin of Species by Means of Natural Selection,* vol. 49 of *Great Books of the Western World,* ed. Robert Maynard Hutchins (Chicago: Encyclopaedia Britannica, 1952), 63.

dition of life and also in most cases to an advance in organization, he was well aware that this evolution does not wipe out lower, simpler forms of life, since they will endure as long as they are well suited for their simple conditions. Darwin's conclusions in terms of the future of species were quite optimistic. Since he thought "natural selection works solely by and for the good of each being," he could claim that "all corporeal and mental endowments will tend to progress towards perfection."[26] But in terms of the modifications, he was not yet convinced that he had found all the causes, since he surmised that "natural selection has been the main but not the exclusive means of modification."[27] While demonstrating a basic unity of life and of its gradual evolution, he was still hesitant to make definite assertions about the actual course evolution took. He thought animals developed from at most four or five progenitors, and plants from an equal or lesser number. But he considered it immaterial as to whether one can really assert that "all the organic beings which have ever lived on this earth may be descended from some one primordial form."[28] Darwin's emphasis in *The Origin of Species* was not so much on an all-embracing evolutionary picture of life. He rather attempted to show that plants and animals are not fixed but tend to develop and show a basic cohesion among themselves.

Considering the groundwork that had been laid for Darwin's ideas, people should have been prepared for *Origin of Species* when it appeared in 1859. Nevertheless, it was met with both eagerness and outrage. The first edition of 1,250 copies sold out on the day of publication, and a second edition of 3,000 copies soon afterward.[29] Even his friend, the British geologist Charles Lyell (1797-1875), pleaded pathetically with Darwin to introduce just a little divine direction into his system of natural selection.[30] Already in March 1860 the American botanist Asa Gray (1810-88) of Harvard University published an extended and careful review of Darwin's book in the *American Journal of Science and Arts*.[31] Gray

26. Charles Darwin, *Origin of Species,* 243.
27. Charles Darwin, *Origin of Species,* 239.
28. Charles Darwin, *Origin of Species,* 241.
29. According to Charles Darwin, *Autobiography,* 122.
30. According to William Irvine, *Apes, Angels, and Victorians: The Story of Darwin, Huxley, and Evolution* (New York: Time, 1963), 130.
31. Asa Gray, "Review of Darwin's Theory on the Origin of Species by Means of Natural Selection," *American Journal of Science and Arts* 79 (March 1860): 153-84, re-

admitted that Darwin's theory of natural selection could be interpreted in atheistic terms. Yet he suggested that such an evaluation could not be arrived at from a scientific basis and reminded his readers that Newton's physics could be interpreted atheistically also. Yet he was convinced that "it is far easier to vindicate a theistic character for the derivative theory [of descent]."[32] In conclusion Gray emphasized once more that Darwin's book is no religious treatise: "The work is a scientific one, rigidly restricted to its direct object; and by its science it must stand or fall."[33]

Although in the first edition of Darwin's treatise one could not notice exactly which standpoint Darwin assumed "theologically speaking," Gray surmised that presumably he did not want to deny a creative intervention in nature. The idea of a natural selection would require so many creative acts which were independent from each other that the whole process must be regarded "more mysterious than ever." Before this review was published Gray had already obtained the second edition of Darwin's book, and he remarked "with pleasure the insertion of an additional motto on the reverse of the title page, directly claiming the theistic view which we have vindicated for the doctrine."[34] At least from the second edition of *The Origin of Species* onward, Darwin emphasized: "There is grandeur in this view of life, with its several powers, having been originally breathed by the Creator into a few forms or into one; and that, whilst this planet has gone cycling on according to the fixed law of gravity, from so simple a beginning endless forms most beautiful and most wonderful have been and are being evolved."[35] Yet he felt that the more he allowed for divine guidance in variations, the less reality would pertain to natural selection.[36]

In the USA Darwinism was received with relative ease and in a theistic gown. Yet it was actually not Darwin and his theory of natural selection which found common acceptance there, but rather the philosophy of the British writer Herbert Spencer (1820-1903) and his cosmic

printed in Asa Gray, *Darwiniana: Essays and Reviews Pertaining to Darwinism,* ed. A. Hunter Dupree (Cambridge: Harvard University Press, 1963), 8-50.

32. Gray, *Darwiniana,* 45.

33. Gray, *Darwiniana,* 49.

34. Gray, *Darwiniana,* 49f.

35. Charles Darwin, *Origin of Species,* 243.

36. Cf. Irvine, 132.

theory of an all-embracing evolutionary process and his thesis of the survival of the fittest.[37] Spencer claimed that an unknown and unknowable absolute power is continuously at work in the material world and brings forth diversity, coherence, integration, specialization, and individuation. For a young and expanding country such as the United States, it was a matter of course that the biological theory of Darwin became an appendix to the social, economic, and philosophical theory of progress as advanced by Spencer. Only the theologian Charles Hodge (1797-1878) of Princeton Theological Seminary challenged Darwinism in his small book *What Is Darwinism?* (1874) by claiming it excluded any final causes and therefore also any plan of a creation. Though he admitted that Darwin believed in a creator, he claimed that his theory is basically atheistic.[38] Along with Hodge, many other Christians in the United States began to show a strong antievolutionary bias. It was not just his own conviction that made Hodge turn against Darwin. He was strongly influenced by the European controversy that surrounded Darwin and his theory of evolution, and feared that this theory might have a similar negative impact in the United States.

In Great Britain there had occurred in Oxford on June 30, 1860, the famous debate between the conservative bishop Samuel Wilberforce (1805-73) and the ardent defender of Darwin Thomas H. Huxley (1825-95). In this context Wilberforce postulated the exaggerated claim that Darwin says: "Humanity is descended from monkeys," a claim Darwin had never made.[39] When in 1871 Darwin also included humanity in his theory of evolution *(The Descent of Man and Selection in Relation to Sex),* he attempted to show that all human characteristics can be explained through gradual modifications of humanlike ancestors through a process of natural selection. Similar to Darwin's first book, here someone else had done spadework, this time the German zoologist Ernst Haeckel in his *The History of Creation* (1868; ET 1876). In his characteristic openness and modesty Darwin remarked about this in his book *The Descent of Man:* "If this work had appeared before my essay had been written, I should probably never have completed it. Almost all the conclusions at

37. Cf. Herbert Spencer, *First Principles* (New York: De Witt Revolving Fund, 1958), where he forcefully sets forth his notion of evolution.

38. Charles Hodge, *Systematic Theology* (Grand Rapids: Eerdmans, 1952), 2:16ff.

39. Cf. Irvine, 5. He "begged to know, was it through his grandfather or his grandmother that he claimed his descent from a monkey?"

which I have arrived I find confirmed by this naturalist, whose knowledge on many points is much fuller than mine."[40] In *The History of Creation* Haeckel, however, went far beyond Darwin. On the one hand we find an optimism which can be hardly understood today. Haeckel, for instance, claims: "We are proud of having so immensely surpassed our lower animal ancestors, and derive from it the consoling assurance that in the future also, mankind as a whole, will follow the glorious career of progressive development, and attain a still higher degree of mental perfection."[41] He refers here especially to Spencer, whose application of evolutionary thought to practical human life opens up "a new road towards moral perfection."[42] But then the religious program of Haeckel comes out when he concludes: "The simple religion of Nature, which grows from a true knowledge of her, and of her inexhaustible store of revelations, will in the future ennoble and perfect the development of mankind far beyond the degree which can possibly be attained under the influence of multi-various religions of the churches of the various nations — religions resting on a blind belief in the vague secrets and mystical revelations of a sacerdotal cast."[43]

In his influential book *The Riddle of the Universe* (1899), Haeckel finally advances with much persuasion a monistic worldview in which the world is considered eternal and infinite. Instead of clinging to the old ideals of God, freedom, and immortality, we should now support "truth, beauty, and virtue," which are "the three goddesses of the monist."[44] The great antithesis between theism and pantheism, vitalism and mechanism is now finally overcome through a "rational knowledge" and a "monistic conception of the unity of God and world."[45] For Haeckel this means that faith in God is no longer useful. "The monism of the cosmos" is now being established which "proclaims the absolute dominion of 'the great eternal iron laws' throughout the universe."[46] Materialism

40. Charles Darwin, *The Descent of Man in Relation to Sex*, vol. 49 of *Great Books of the Western World*, 254.

41. Ernst Haeckel, *Natürliche Schöpfungsgeschichte*, 4th ed. (Berlin: Georg Reimer, 1873), 656.

42. Haeckel, *Natürliche Schöpfungsgeschichte*, 657.

43. Haeckel, *Natürliche Schöpfungsgeschichte*, 658.

44. Haeckel, *Riddle of the Universe*, 336.

45. Haeckel, *Riddle of the Universe*, 290.

46. Haeckel, *Riddle of the Universe*, 381.

has been victorious in the dogma of evolution, and faith in God was in retreat.

Today the optimism of the nineteenth century has largely been shattered. Scientific knowledge has led us not only to new heights but also to existential threats to life, which appeared exactly when we were convinced that the future was unconditionally open to us. The American theologian Langdon Gilkey (b. 1919) pointed out at a conference of Nobel laureates: "It has been another cherished myth of our culture that technology raises only technical problems and that to every technical problem there is in potentiality a technological answer."[47] He pointed out that power and knowledge also blind us and can lead to self-destruction. Therefore he called us to repentance and to a new humility in our scientific and technological culture, because "the future that science brings to us — as well as the future of science itself — may well be darkness and not light."[48] An indication that science has regained a certain degree of modesty can be seen not only in the fact that instead of reality it only talks about models of reality, but also in the increasing number of scientists who become suspicious of the decided exclusion of the reference to God. For instance, some twenty years ago the American astronomer and agnostic Robert Jastrow (b. 1925) wrote: "For the scientist who has lived by his faith in the power of reason, the story ends like a bad dream. He has scaled the mountains of ignorance; he is about to conquer the highest peak; as he pulls himself over the final rock, he is greeted by a band of theologians who have been sitting there for centuries."[49]

47. Langdon Gilkey, "The Future of Science," in *The Future of Science: 1975 Nobel Conference,* ed. Timothy C. L. Robinson (New York: John Wiley, 1977), 124.

48. Gilkey, 125.

49. Robert Jastrow, *God and the Astronomers* (New York: W. W. Norton, 1978), 116.

Part II
The World in Scientific Perspective

On account of immense technological advances, our life has changed more during the last century than in all the previous centuries combined. Human life expectancy has nearly doubled, our pace in covering distances has multiplied several times, and the living standards of many people in central Europe and North America have increased to a level never before envisioned. This rapid technological change can be traced back to fundamental discoveries in the sciences. Many laws and theories that had been considered virtually sacrosanct were either completely reformulated or discarded. In the twentieth century we arrived at a new understanding of the world which is largely shaped by momentous advances in the sciences and which we will now describe in broad outline. We will highlight this new worldview as it pertains to our understanding of the universe, life, and the causal nexus.

1. The Universe

For many people the fundamental difference between the modern worldview and that of the nineteenth century is best illustrated in modern cosmology.[1] The fundamental concepts of the world were irrevocably shaken. Representative of the new view is the work of Albert Einstein (1879-1955). Having shown the close connection between space, time, and matter, he asked, like many others, what consequences these theoretical considerations have for our understanding of the universe. He especially questioned whether the traditional assumptions concerning the age of our universe and its extension in space and time could still be valid. In 1916, in putting forth his general theory of relativity, Einstein partially answered these questions. He suggested that gravitational fields produced by masses distributed in space and moving in time are equivalent to the curvature of the four-dimensional space-time continuum which they form. If applied to the forces of gravity, this proposal would mean that the light emitted from distant stars, just grazing the sun, is deflected somewhat toward the sun when it passes by the sun and its immense gravitational pull. This curvature is high enough that it can be verified by observation. For instance, if two stars have an angular distance slightly larger than the angular diameter of the sun, it can be observed that the angular distance between these two stars is slightly larger when the sun has "moved away" from this part of the sky than

1. A very readable introduction to the various cosmological theories is provided by Norriss S. Hetherington, ed., *Cosmology: Historical, Literary, Philosophical, Religious, and Scientific Perspectives* (New York: Garland, 1993). Quite instructive is the volume *Modern Cosmology in Retrospect,* ed. B. Bertotti et al. (Cambridge: University Press, 1990), with contributions by Hoyle, Bondi, and others.

when it comes to stand directly between the two stars. Of course, one needs the help of a total solar eclipse to measure the angle between the two while the sun stands between them. Indeed, such a change in the angular distance has been experimentally proven.

Einstein attempted to generalize the gravitational pull by applying it to the structure of the whole universe. He asserted that the whole space-time continuum associated with the mass of matter contained in our universe is bent and curved. For a while the question remained unsolved as to whether the curvature is positive or negative: a positive curvature would mean that the universe is finite; a negative curvature would prove the infinity of the universe. It was relatively easy to decide that the curvature is not varied like that of a banana but uniform like that of a ball, because the galaxies seem to be evenly dispersed throughout the observable universe. But to decide between a negative and a positive curvature on the grounds of the observable phenomena is very difficult. Einstein suggested that the curvature was positive and thus the universe finite. However, he held the curvature of space to be independent of time and consequently conceived of a space-time continuum similar to a cylinder with the time axis running parallel to the axis of the cylinder and the space axis perpendicular to it.

In 1917 Willem de Sitter (1872-1934), a Dutch mathematician, showed much more convincingly that both space and time must be curved (which seems more logical according to the presuppositions in Einstein's special theory of relativity), and in turn suggested an expanding universe, similar to an expanding globe with a longitude serving as the space coordinate and a latitude as the time coordinate. In this model of the universe, "history" would not repeat itself, unlike the model proposed by Einstein. For instance, a light ray would not travel in a complete circle, but in a perpetually expanding spiral. In 1922 the Russian mathematician Alexander Friedmann (1888-1925) applied Einstein's theory to the universe as a whole and proved there were no static solutions to account for the structure of the universe. Einstein, however, was horrified about this proposal and in turn introduced a cosmological constant according to which the universe would neither expand nor contract. Only much later, after the American astronomer Edwin Hubble (1889-1953) discovered that the light coming from distant galaxies is subjected to a redshift in its spectral lines which increases with the respective distance of the galaxies from us, did Einstein admit that he had made "the biggest blunder of his life" and correct the shortcoming of his own proposal.[2] This redshift is analogous to the "distorted" sound of a car that passes us on the road and moves away from us. The larger the redshift, the faster the illuminated object moves away from us.

The universe seems to be in a state of uniform expansion whereby the mutual recession velocities between any two galaxies in space are proportional to the distance between them. It is interesting here that Friedmann's proposal is not irreconcilable with Einstein's theory. Classical Newtonian mechanics also did not have static solutions for interpreting the structure of the universe. According to Newtonian mechanics, if at any given moment the galaxies had been static, they would certainly have to start moving toward each other under their mutual gravitational attraction. This would lead to a contracting universe collapsing into itself. The only alternative would be for the universe to expand so the galaxies would have enough residual velocity to move away from each other against the contrary gravitational drag. We can demonstrate this with a ball that we toss up in the air. This ball does not remain standing motionless in the air. It continues its upward move, away from the earth and its gravitational field, until its motion comes to a halt, and then it starts falling. It is amazing that the idea of an unchanging eternal universe still exerted such fascination for an innovative scientist like Einstein in the early decades of the twentieth century. Yet, together with Friedmann, he laid the foundation with his mathematical work for the so-called big bang theory.

a. Theories concerning the Beginning of the Universe

The Belgian astronomer Georges Lemaître (1894-1966) took as his starting point de Sitter's version of the Einsteinian model of a universe and in 1925 postulated a continuously expanding universe and then projected it back to its origin. He assumed that if the universe is continuously expanding, it must have had a beginning or a zero point at which

2. According to Martin Rees, Remo Ruffini, and John A. Wheeler, *Black Holes, Gravitational Waves, and Cosmology: An Introduction to Current Research* (New York: Gordon and Breach, 1974), 153. For an extensive discussion of the shift from a static to an expanding universe, cf. Robert W. Smith, *The Expanding Universe: Astronomy's "Great Debate," 1900-1931* (Cambridge: University Press, 1982).

it had almost no extension. This zero point of cosmic time must date further back than the age of the oldest surface rocks on our earth and the age of meteorites. It can be estimated by retracing the most distant galaxies to a state where they were all assembled at the same starting point. These different calculations converge upon approximately the same date. Various investigations have shown that the age of the oldest surface rocks is roughly 3.5 billion years, that of meteorites 4.7 billion, and the big bang model would yield little less than 10 billion; through conclusions gained from cosmic background radiation data, one model updated the age of the universe to approximately 15 billion years.[3] This means that roughly 15 billion years ago the mass of the universe must have been so tightly packed together that no individual atoms could exist, but only, as Lemaître called it, a "cosmic egg" or a "primeval atom" of extraordinarily high radiation.[4] Lemaître concluded that this primeval nucleus of immense proportions broke with a burst to form the atoms we know today. Initially very massive atoms were formed, breaking down further to smaller atoms until one arrived at stable chemical elements such as oxygen, helium, carbon, and iron. If one accepts the hypothesis of an initial big bang, then these conclusions sound quite convincing. Yet the atomic structure of the universe deduced from this hypothesis would consist mainly of heavy elements. In fact, however, our universe consists of 90 percent hydrogen and 9 percent helium, while the heavier elements account for only 1 percent.

In 1948 the Russian-American astrophysicist George Gamow (1904-68), together with Hans Bethe (b. 1906) and Ralph Alpher (b. 1921), presented a different approach to the big bang theory. He claimed that the cosmic egg was so highly compressed that it did not contain individual atoms but only "ylem," a very hot primordial gas. This ylem, a term Alpher coined, consisted of protons and electrons packed together so tightly that they formed a mass of electrically uncharged particles called neutrons. At an early stage of the expansion, approximately "five minutes" after the big bang, the universe had already cooled down sufficiently to allow the aggregation of protons and

3. Cf. James Trefil, *The Moment of Creation: Big Bang Physics from Before the First Millisecond to the Present Universe* (New York: Scribner, 1983), 24, who provides a very readable account of the origin of the universe.

4. Cf. Georges Lemaître, *The Primeval Atom: An Essay on Cosmogony,* trans. Betty H. Korff and Serge A. Korff (New York: D. van Nostrand, 1950), 126f.

neutrons into complex nuclei, namely, deuterons (H_2), tritons (H_3), tralphas (He_3), alphas (He_4), and others. Yet within "thirty minutes" of the big bang, the temperature would have been below that required to sustain thermonuclear reactions in light elements. Furthermore, free neutrons, which were abundant at the beginning but have a half-life of only thirteen minutes, must have virtually disappeared. They were either used up in the formation of elements or decayed to protons to form the nuclei of hydrogen atoms.

Physicists Enrico Fermi (1901-54) and Anthony Turkevich (b. 1916) studied the thermonuclear reactions Alpher, Bethe, and Gamow had postulated and came to the conclusion that if the ylem hypothesis was correct, it would have resulted in an equal amount of hydrogen and helium and about 1 percent of deuterium from which the heavier elements could have been formed by capturing neutrons. However, Gamow himself noticed that his theory could account only for the formation of hydrogen and helium, but not for the existence of heavier elements except in very unlikely circumstances, because once the atomic mass of four (helium) is reached, a simple addition or a capturing of a neutron or a proton leads to extremely unstable elements. Since there is no stable nucleus with atomic mass five, the amount of heavier elements, calculated on the basis of this theory, would be much lower than the amount actually observed.

The British astronomer Fred Hoyle (b. 1915) provided some assistance here. He suggested that hydrogen is the only original material. Everything else is formed within stars and added to the interstellar material.[5] Supernova luminosity decays over a period of fifty-five days. Hoyle along with the American nuclear physicist William Fowler (b. 1911) discovered that this period coincides with the natural decay period of the spontaneous fission of californium 254. Thus they concluded that large amounts of californium are formed in the process of a supernova explosion. Furthermore, it had been observed that the spectrum of certain stars shows the presence of technetium. This element is radioactive and possesses no stable variety. The most stable form is technetium 99 with a half-life of about 220,000 years. If we allow 5 billion years for the average life of a star, this would mean that only one-billionth of the originally present amount of technetium 99 would remain. Yet there is still suffi-

---

5. Fred Hoyle, *The Nature of the Universe* (New York: Harper, 1950), 85.

cient technetium in the stars to show in the spectral lines of their light. Two possibilities could account for this phenomenon. Either technetium was initially present in rather huge amounts, which would be rather unlikely, or it is being formed in the interior of the stars. The latter possibility would support Gamow's thesis of the formation of the chemical elements. If Hoyle's thesis is true (that this process goes on in the stellar core, where the density of matter would be so much higher than in open space), this would point to the origin of heavier elements. The odds of a helium 4 nucleus being struck by two particles essentially simultaneously would be much better in stars. Thus heavier atoms that lead to stable varieties could be more easily formed.

By 1970, however, this idea of a nucleosynthesis in stellar bodies through which the heavier elements were formed was gradually abandoned in favor of an explosive nucleosynthesis, such as in supernovas, through which a star explodes to end its life.[6] The present cosmological theories of a big bang assume that during the first minutes after the big bang no atomic nuclei were formed because of the extreme heat. Once the temperature had cooled to 10,000,000 degrees, protons and neutrons came together to form heavy hydrogen (deuterium) as well as helium He_3 and He_4. The result was that 76 percent of the matter remained as single protons which then later connected with an electron, each forming common hydrogen. Twenty-four percent of the matter turned into helium nuclei (He_4), and then to some extent lithium and deuterium as well as He_3 were formed — that today there is 24 percent helium in the universe is not just a consequence of this theory, but also a test to its truthfulness, as far as we can determine at the present point. During the next one thousand years, the universe underwent considerable changes. Initially cosmic radiation dominated, but then the universe filled with matter. After it cooled down galaxies were formed from interstellar matter. As these galaxies condensed more and more, they heated up again, as did intergalactic matter. The steadily decreasing density of intergalactic matter barely allowed for condensations. Therefore we can assume that all galaxies were formed at approximately the same time after the big bang, roughly 8 to 10 billion years ago, in the subsequent process of expansion. The first stars in these galaxies con-

6. Cf. W. David Arnett and Donald D. Clayton, "Explosive Nucleosynthesis in Stars," *Nature* 227 (August 22, 1970): 780-84.

sisted of hydrogen and helium. Since the first generation of stars were much larger in mass than our sun, they often ended their life with explosions similar to those of supernovas. Because of these explosions, many of the heavier elements were formed. It has even been discovered that elements are formed subsequent to such explosions. For instance, in the region of Orion, in which there has been supernova activity in more recent times, there are streams of carbon and oxygen nuclei with weak energy, streams thirty times stronger than those closer to our sun. This process of the development of heavy atoms therefore could also take place at other "birthplaces" of stars in our Milky Way.[7]

Three decades of scientific research have unearthed surprising details which have added strength to the theory of an expanding universe and considerably improved it in many details. For instance, in 1948 the theoretical physicists Gamow, Alpher, and Robert Hermann postulated that the universe would have cooled down after 15 billion years to a temperature of a little more than 3 degrees Kelvin, that is, somewhat higher than absolute zero. This temperature was proven to exist through the discovery of a cosmic background radiation by Arno A. Penzias (b. 1933) and Robert W. Wilson (b. 1936), who in 1965 calculated that the temperature was a little lower than originally postulated.[8] They found that at a wavelength of 7.35 cm there is considerable background noise which is independent of any direction and is present day and night. This noise could neither come from the Milky Way, since it does not correspond to any direction, nor be induced by solar radiation. The electromagnetic radiation which caused this noise was found to be black body radiation. It was not dependent on any material — only on temperature — and could be correlated according to its wavelength with a temperature of 2.7 degrees Kelvin, which is -270.3 degrees Celsius. This means that our earth is subjected to an electromagnetic radiation which can be considered a reverberation "of the explosion with which the universe began."[9] Using known physical laws, science has come as close as it can "to the moment of creation itself."

Our knowledge of the beginning extends back to 10^{-35} or even 10^{-43}

7. Cf. to this R. Cowen, "Supernovas Help Solve an Elemental Mystery," *Science News* 147 (February 5, 1995): 70.

8. Cf. R. B. Partridge, *3 K: The Cosmic Microwave Background Radiation* (Cambridge: University Press, 1995), 47f.

9. Trefil, 20, for this and the following quote.

seconds after the big bang occurred. If the explosion of a cosmic egg is the beginning of the universe, then the question arises as to what was before the so-called zero point. Of course, we could conjecture that the once stable cosmic egg suddenly or gradually became unstable. Yet the question which then emerges is what made it change from its once stable state to an unstable one. If the big bang theory is assumed to be true, the dilemma is often solved by saying there was a state prior to the cosmic egg where matter/energy existed in a form similar to the intergalactic matter that we experience today.

This suggests the existence of an exceedingly thin gas which is subject to its own vastly diffused gravitational field. Slowly gas collected and the universe drew in upon itself. As the substance of the universe grew more compact, the gravitational field became more and more intense until, like a snowball rolling down a hill, matter contracted at an ever increasing rate. Since matter was compressed into a smaller and smaller volume, the universe heated up. This heat increase countered the gravitational pull and the contraction began to slow down. The inertia of matter, however, kept it contracting, passing the point where temperature and gravitation were in balance. The universe approached its minimum volume represented by the cosmic egg or the ylem state of matter. Finally the temperature was so high and the radiation so intense that the outward pressure due to the hot gases exceeded the inward pressure due to gravity. The substance of the universe was pushed out faster and faster and the big bang occurred. The universe had a definite beginning: the immensely thin gas which went through a stage of condensation and then, after the big bang, subsequently expanded.

b. Theories concerning the Future of the Universe

It is difficult to obtain a satisfying answer from science about the future of the universe. Following the theory of a constantly expanding universe, the decisive question is whether the velocities with which the galaxies recede from each other are large enough for mutual escape. The problem can be illustrated by throwing a ball in the air. If we throw it with an initial velocity that is high enough, the ball escapes into space and does not return in spite of the ever present force of gravity. If the velocity is not high enough, then the initial velocity will decrease to zero and increase again in reverse until the ball hits the earth. Many scientists indeed suggest that the first possibility bears closer resemblance to the future of our universe. The recessional velocity of the galaxies is believed to be high enough that they will continue to expand in spite of the gravitational attraction between them. Thus astronomers speak of a hyperbolic structure of the universe or of an ever expanding universe. Every galaxy outside our own will continue to recede with increasing velocity, corresponding to the increasing distance between the galaxies and the diminishing gravitational attraction. Seen from our vantage point, the galaxies will grow dimmer and dimmer and will finally approach the limit of the observational universe. The universe would continuously expand and, so to speak, empty itself more and more. At the end we would be alone in the universe, and all the "celestial lights" would be extinguished.

Moreover, our own solar system as well as the Milky Way are subjected to an aging process. First the sun will become a red giant and its heat and radiation will increase many times; finally it will turn into a white dwarf and our planetary system will become ultimately uninhabitable. This scenario ties us to an irreversible timeline. There is no rejuvenation because our universe, the Milky Way, our solar system, and earth are subject to an inescapable aging process. Philosophers and theologians are not the only ones who dislike this kind of perspective for the future. Cosmologists too have developed other models which contain more positive aspects and attempt to determine whether these models are more probable.

In contrast to the hypothesis of a continuously expanding universe, one could also support the thesis that this expansion will decrease at some point and, after reaching a zero point of expansion, will reverse and begin to contract. The galaxies would be similar to the extremely thin gas which was dispersed throughout the universe prior to the cosmic egg and condensed itself and subsequently exploded. Since the first law of thermodynamics, or the so-called law of the conservation of energy, states that in a closed system no energy can get lost, the fluctuations between expansion and contraction could continue almost indefinitely. This conclusion leads to the theory of a pulsating universe. Yet one cannot make any assertions concerning the laws and structures of any subsequent or preceding cycles, and thus each cycle would be singular.

Matter and Antimatter Universe

If one assumes that our universe is paired with an "anti-universe," one could also arrive at a pulsating system. In 1929 the English physicist Paul Dirac (1902-84) postulated that for each type of particle there exists an "antiparticle" with the exact opposite characteristics. Three years later, in 1932, the American physicist Carl D. Anderson (b. 1905), in studying a cosmic radiation, discovered the antiparticle to the electron, which was first called an antielectron and later on a positron. However, it was not until 1955 that scientists working with the Bevatron accelerator at the University of California in Berkeley produced and observed the first antiproton. It took another ten years until physicists at the Brookhaven National Laboratory in Upton, New York, succeeded in building an artificial antinucleus by combining an antiproton and an antineutron to an antideuteron, the counterpart of the nucleus of ordinary heavy hydrogen. Encouraged by these discoveries, some scientists, such as the American physicist Maurice Goldhaber (b. 1911), suggested that "since neither a particle nor its antiparticle can have a preferred position," there must also be an anti-universe consisting of antimatter.[10] From this hypothesis one can arrive at a pulsating universe, in which universe and anti-universe merge in the process of contraction whereby the masses of both are converted into energy according to the formula $e = mc^2$ (energy equals mass times the square of the speed of light) and form a cosmic egg of gamma rays.[11] Perhaps the immense radiation pressure of these energy-rich photons is sufficient to account for a big bang with a subsequent expansion through which particles and antiparticles are formed and originate their own "worlds." The American science fiction writer and scientist Isaac Asimov (1920-92) finally suggested that universe and anti-universe are similar to two connecting balloons: while one universe contracts the other expands, and vice versa. Thus the whole system would be static while in relation to each other; the universes are in a continuous process of expansion and contraction.

The obvious symmetry between matter and antimatter is very sug-

10. Maurice Goldhaber, "Speculations on Cosmogony," *Science* 124 (August 3, 1956): 218f.

11. Cf. Hannes Alfvén, *Worlds-Antiworlds: Antimatter in Cosmology* (San Francisco: W. H. Freeman, 1966), 101f.

gestive of the hypothesis of a universe and an anti-universe. Yet matter and antimatter in close proximity will annihilate each other. No star could contain a close mixture of matter and antimatter. Otherwise it would explode more violently than a supernova. Observation of cosmic radiation has not led to the detection of sufficient gamma radiation which points to as much matter-antimatter annihilation as would be produced if interstellar or intergalactic gas were to contain large amounts of antimatter. Even to suggest that matter and antimatter must be attributed to respective galaxies leaves us with the question of how matter and antimatter could be neatly separated in respective galaxies if they were created together. As far as we know, the gravitational interaction between matter and antimatter is identical to that between matter and matter, suggesting an indiscriminate mixture rather than a distant segregation. Also abandoned is the assumption that all the vast energy releases observed in quasi–stellar objects, commonly called quasars, could be explained in terms of annihilation of matter and antimatter.[12] Quasars without measurable radiation have even been discovered. We may, however, concede that the presence of large amounts of antimatter in the universe cannot be ruled out completely, nor can we entirely reject the idea that some cosmic sources of intense radiation might indeed be due to the annihilation of matter and antimatter.

Regardless of these concessions, it is unlikely that the total amount of antimatter in the universe is anywhere close to the fifty-fifty percentage that a matter-antimatter cosmology assumes. Perhaps we should follow Hannes Alfvén (b. 1908), who advocates the existence of matter and antimatter galaxies, yet cautions that any statement about the existence of equal amounts of matter and antimatter in the universe has to be founded on an artificial assumption that is not subject to observational tests.[13] If the hypothesis of a matter-antimatter "creation" of the universe could be verified, it would however not necessarily overrule the depressing thought of an irreversible aging process of the world. Instead of start-

12. Geoffrey Burbidge and Margaret Burbidge, *Quasi–Stellar Objects* (San Francisco: W. H. Freeman, 1967), 210ff.

13. Cf. Hannes Alfvén, "Antimatter and Cosmology," *Scientific American* 216 (April 1967): 106-14, who cautiously postulates the existence of galaxies of antimatter and others of matter. One should not overlook the caution, however, that is expressed against Alfvén's theories. Cf. Helge Kragh, *Cosmology and Controversy: The Historical Development of Two Theories of the Universe* (Princeton: Princeton University Press, 1996), 383f.

ing with a primordial fireball of ylem, as in the big bang theory, one would start with ambi-plasma containing both particles and antiparticles. Through initial gravitational contraction and subsequent radiation pressure, due to proton-antiproton annihilation, we would then encounter an expanding universe, which we still observe with our telescopes.

Continuous Creation (Steady State Theory)

A very different approach toward cosmogony was taken by the proponents of the steady state theory or theory of continuous creation. To some extent this theory seems to return to Einstein's earlier model of a stationary universe. The founders of the steady state theory were sympathetic to Einstein's original model of an eternal and unchanging universe. They rejected the idea that matter could have been created at some point when the universe began to exist. Three English astronomers, Herman Bondi (b. 1919), Thomas Gold (b. 1920), and Fred Hoyle, introduced the hypothesis that matter has no beginning but is continuously being created even today. To underline this idea they advanced in 1948 the so-called perfect cosmological principle, according to which the universe appears the same at all times and in all places.

Two main objections seemed to contradict this assertion. First, we remember that galaxies are constantly receding from each other and thus our universe becomes more and more dispersed. Second, we have heard that in stars, such as our sun, hydrogen is steadily being transformed into helium, and therefore the hydrogen content of the universe is constantly decreasing. Bondi, Gold, and Hoyle tried to overcome these observations that seem to contradict their hypothesis by reformulating the law of conservation of energy. They said that, contrary to common opinion, this law does not suggest that energy is never created out of nothingness but only states that energy has never been observed to be created out of nothingness. They now claimed that such a creation, a continuous creation, actually takes place and compensates for the loss of energy of the receding galaxies and for the production of helium. The continuous creation they suggested is so minute that only one hydrogen atom is created per year in 1 billion liters of space. Such a profoundly small amount suffices to balance the hydrogen loss, but is still much too small to be ever discovered.

Since the perfect cosmological principle does not allow room for a big bang or a cosmic egg, the recession of the galaxies must have a different origin. Hoyle, for instance, claimed that "the new material produces an outward pressure that leads to the steady expansion," while the British astronomer Raymond A. Lyttleton suggested that the positive charge of the proton is slightly stronger than the negative charge of the electron.[14] Though the difference of these charges is much too small to be detected through direct measurements, it would allow for a strong enough amount of positive charges in galaxies to be built up so that they undergo a continuous mutual recession similar to the repulsion of the same poles of two magnets. The steady state theory can do without a creation in the beginning, since it suggests a self-rejuvenating universe that has no beginning and no end. Our universe always has been and always will be the same. Hoyle himself sees the essential difference between the steady state theory and the former theories like this: "Without continuous creation the Universe must evolve toward a dead state in which all the matter is condensed into a vast number of dead stars. . . . With continuous creation, on the other hand, the Universe has an infinite future in which all its present very large-scale features will be preserved."[15] Small wonder that, confronted with this kind of "eternal version," Hoyle has not much positive to say about the Christian understanding of eternal life.[16]

Yet the steady state theory itself is not as well founded as its proponents first believed. To start with, one wonders by which principles certain physical laws are picked out for modification while others are left unaltered. It almost seems that there are certain a priori principles at work that are subsequently applied to reality. However, it was a better understanding of reality that made the steady state theory falter. Since this theory was first proposed, several hundred quasi-stellar objects, or quasars, have been discovered and observed individually that emit microwaves, and some even visible light. Another estimated 1 million quasars have become accessible through large radio telescopes. The emission lines of these quasars show a unique redshift that seems to indicate

---

14. Hoyle, 126; and cf. Raymond A. Lyttleton, "An Electric Universe?" in *Rival Theories of Cosmology: A Symposium and Discussion of Modern Theories of the Structure of the Universe,* ed. H. Bondi, W. B. Bonnor, R. A. Lyttleton, and G. J. Whitrow (London: Oxford University Press, 1960), 25ff.

15. Hoyle, 132.

16. Hoyle, 139f.

that these objects are receding from us with velocities of up to 80 percent of the speed of light — several times higher than those of usual galaxies. The farther these objects are away from us, the larger the redshift. This would mean that these quasars are all very far away from us and therefore must have been formed many billions of years ago.[17] Since none of these objects has been discovered in our more immediate environment, the universe does not seem to be as uniform as steady state theory proponents assert. The process through which quasars originate no longer seems to be operative in our vicinity. Therefore the conclusion is unavoidable that the universe changes its appearance and goes through an aging process. When cosmic background radiation was discovered in the 1960s, the steady state theory lost its credibility altogether. Yet there are still other possibilities to explain our universe without the hypothesis of a big bang.

Theses from the Quantum Cosmology

The development of the quantum theory demonstrated that, scientifically speaking, one cannot start a theory from the point $t = 0$ (i.e., the very beginning of the big bang), since the physical state at this point is not defined. The conditions at this point were extreme: density, temperature, and pressure were almost infinite. Importantly, physical laws governing at these extreme conditions are unknown. The first assertions can be made after the so-called Planck time (i.e., once an interval of 10^{-43} seconds after the beginning of the universe has elapsed). This interval is called the quantum cosmos. The development of the universe during the subsequent period between 10^{-43} and 10^{-35} seconds is called the era of the grand unified theory.

Within the realm of the grand unified theory one attempts to trace the events at the early stages of our universe through various hypotheses. The most interesting attempt comes from Stephen Hawking (b. 1942) of

---

17. Cf. Maarten Schmidt and Francis Bello, "The Evolution of Quasars," *Scientific American* 224 (May 1971): 54-69. As the number of quasars increases up to a certain point with increasing redshift, one could assume that quasars were relatively dense in the universe in its younger age, while they move now toward the "border" of the universe. Cf. also Kragh, 332ff., for the discovery of quasars and the implications for the steady state theory.

Cambridge University. He is considered by some the most brilliant theoretical physicist since Einstein, and he occupies the same chair that was once held by Isaac Newton. Hawking starts with a universe without a singularity. Singularity here means the beginning of the universe, that state by which the laws of physics break down and one cannot make precise assertions. This breakdown occurs because of the extreme density and temperature present at the beginning. Hawking is aware that we do not have accurate dates of observation with regard to the earliest phase of the universe and that therefore his thesis of a universe without an initial starting point cannot be proven. Nevertheless, he proposes that the universe has no borders in space and time and therefore is causally closed. He points out that the theory of relativity only refers to the macro-universe. Therefore it cannot adequately describe the initial problems of the universe. If the original singularity was very small and dense, then we must resort to another theory: quantum mechanics. The theory of relativity and quantum mechanics are unified to a quantum theory of gravitation. Therefore one can describe a universe with a single mathematical model in which Hawking also includes Heisenberg's uncertainty relation, which says, among other things, that the subsequent course is not defined by the original boundary conditions.

For space and time, according to Hawking, there is no limit determination altogether, since the curved space-time dimension which limits itself does not lead to an absolute point zero prior to which there was no time. According to Hawking, "it is possible for space-time to be finite in extent and yet to have no singularities that formed a boundary or edge. Space-time would be like the surface of the earth, only with two more dimensions."[18] Hawking immediately draws theological conclusions from his universe without a singularity: "There would be no singularities at which the laws of science broke down and no edge of space-time at which one would have to appeal to God or some new law to set the boundary conditions for space-time. One could simply say: 'The boundary condition of the universe is that it has no boundary.' The universe would be completely self-contained and not affected by anything outside itself. It would neither be created nor destroyed. It would just be."[19]

---

18. Stephen Hawking, *A Brief History of Time: From the Big Bang to Black Holes* (New York: Bantam Books, 1988), 135.
19. Hawking, 136.

In many ways the theory of a universe without a singularity resembles the steady state theory. There, too, one starts with an original given which never changes much. Similar to the steady state theory, the theses of Hawking were received with decided rejection by some and with strong agreement by others. The reason for this was his blunt question: "What place, then, for a Creator?"[20] God's freedom to act is extremely limited by a comprehensive system of laws. Within this exhaustive scientific description of the world, God as the provider of an initial singularity has lost the last niche in which he could meaningfully survive.

But we should not limit our understanding of God's creative activity simply to the inauguration of a beginning, since God is also with the creation. The creative act at the beginning only represents the first act of God's creation; it is not the whole of God's work. For instance, on the basis of its experience of the God who was with them in Egypt, who accompanied them on their journeys and who finally gave them the Promised Land, Israel concluded that this God alone could have created the world. Belief in God the creator is not the starting point of faith, but the result so to speak of a concluding reflection about one's own faith in God. The thesis of a universe without a singularity does not eliminate the "mystery" of a beginning. Hawking simply assumes that space and time have always existed as we see them today. When we consider other theories that are more oriented toward an initial starting point, we notice that the laws of nature become more uncertain the closer we move to the initial state. Finally, at the extreme density and temperature of matter/energy at this initial starting point, they collapse completely. There is no physical theory that could explain without contradiction how the universe existed at the point $t = 0$, since there exists no definition of a physical state at that point. Moreover, we should not become so entranced with the question of a beginning, regardless of whether we consider the universe as emerging from a zero point or as having always been. The universe, at least in those parts we can observe, moves irrevocably toward an "end."

---

20. Hawking, 141, and cf. Keith Ward, *God, Chance, and Necessity* (Oxford: Oneworld, 1996), 43f.

A Dying Universe?

Along with the first law of thermodynamics or the law of the conservation of energy we must also take heed of the second law of thermodynamics, sometimes known as the law of entropy, which states that in an isolated system the entropy or nonconvertibility of energy never decreases but either remains constant or increases. In contrast to the opinion of many materialists of the nineteenth century, our world is no *perpetuum* mobile, an arrangement which continuously keeps running. Rather it resembles an automobile which at one point will run out of gas and then coast a little until its forward movement comes to a final halt. Of course, in a movie we can reverse any given process. Yet the amused reaction of the observers tells us that such events do not occur in reality. Similar to a watch running only as long as there is some resilience in its spring or some energy contained in its battery, all processes within our universe move toward a point at which they will stop. Therefore some scientists talk about a "time arrow" that bars events from being repeated.[21] Considering the size of our universe and the exact maintenance of the planetary orbits, it is difficult for us to understand that all movements of these sidereal bodies are singular. Yet the interstellar gas in our universe will slow down all the heavenly bodies until their kinetic energy is fully used up. Similar to the spring of a huge clock which is totally unwound, our universe will become more and more homogeneous until it simply drifts and there are no discernible changes going on within it.

But there is the possibility that our universe will not keep on expanding until it has reached the state of maximum entropy. The expansion could be changed into a gradual contraction of the universe. Similar to the gravitational collapse of a star, the velocity of contraction of our universe will continue to increase and matter and energy will be compressed again together into an immensely compact fireball. More recent cosmologies interpret this fireball in analogy to a black hole which emerges at the final end of some stars. If a star is considerably larger than our sun, its final collapse will not stop at the stage of a white

21. Arthur S. Eddington, *The Nature of the Physical World: The Gifford Lectures, 1927* (New York: Macmillan, 1929), 68ff., seems to have used the term "time's arrow" for the first time.

dwarf. Under the impact of its own gravity it will collapse to such an extent that a black hole emerges which is so dense that it draws everything into it. Nothing can escape from it, not even light. The only item an observer can measure from the outside is its mass, its electric load, and its torque. This black hole, which consists of almost infinitely dense mass, has such a large gravitation that it cannot be observed in itself apart from its influence on its environment.[22]

Transposed on the universe, this would mean that the collapse of the universe into a gigantic black hole is followed by a subsequent explosion and a new expansion. One could conclude that our universe therefore always moves between a stage of expansion and one of contraction. The total phenomenon of this oscillation would nevertheless be subject to the law of entropy. The movements of expansion and contraction would slowly decrease and finally stop altogether. Of course, one could object that each new cycle would be determined by a totally new set of laws. Under these presuppositions we would have no possibility whatsoever to estimate what such a new cycle would be like. We could at most project an uncertain and unknown future, perhaps of infinite duration, whereby each new cycle would totally annihilate what had come previously. This would mean that we cannot trace back the history of our universe beyond this black hole out of which our present universe emerged. According to this theory, the beginning and the end of our universe would be shrouded in an impenetrable mystery.

We could follow, of course, the French Jesuit and paleontologist Pierre Teilhard de Chardin (1881-1955), who claimed that entropy is perhaps sufficient to determine the future of inanimate nature. But it cannot be transposed to life.[23] According to Teilhard, life shows in each moment that it progresses toward greater complexity and diversity. Through its very success it cannot be subjected to physical entropy. There cannot be a total death of the animate world, since the stream of life is irreversible. Such an argument may at first be comforting when we are confronted with an ultimate and complete balance of all levels of all energy, which is predicted according to the law of entropy. Yet we should not overlook that life can only be maintained by drawing on the

22. John Wheeler coined the expression "black hole" in 1969.
23. Cf. Pierre Teilhard de Chardin, *The Vision of the Past*, trans. J. M. Cohen (New York: Harper, 1966), 168ff.

resources of the inanimate world. What kind of future would we have if all the natural resources are exploited and the sun no longer provides its light through which life can thrive and grow? We cannot separate life from the context in which it is embedded. Though it may be uncomfortable and disturbing, we must admit that within our world there is no eternal life force. The world in which we now live moves irrevocably toward its death.

2. Life

Our understanding of life is deeply influenced by Darwin's theory of evolution and the scientific and philosophical presuppositions connected with it.[1] When Charles Darwin introduced his theory of descent, he already warned: "Our ignorance of the laws of variation is profound. Not in one case out of a hundred can we pretend to assign any reason why this or that part differs, more or less, from the same part of the parents."[2] Though Darwin himself was deeply convinced of evolutionary progress, his theory is still tainted by the mystery of spontaneous variations. Why do such variations occur and how do they affect his thesis? Darwin could give only very imprecise answers. Yet this did not stem the tide of evolutionary thoughts.

a. The Genetic Foundation of Evolution

Due in part to rapid industrialization, the idea of progress and evolution largely dominated the second half of the nineteenth century. Yet it still lacked a scientific basis. Here the work of the Augustinian monk Gregor Mendel (1822-84) was revolutionary and sobering at the same time. In the garden of his Augustinian monastery in Brünn (Bohemia), he had been working for eight years crossing certain kinds of peas, beans, and

1. For a readable and instructive survey of the origin of life, cf. Douglas J. Futuyma, *Evolutionary Biology,* 3rd ed. (Sunderland, Mass.: Sinauer, 1998).
2. Charles Darwin, *The Origin of Species* (1859), in *The Works of Charles Darwin,* ed. Paul H. Barrett and R. B. Freemann (New York: New York University Press, 1988), 15:120.

hyacinths. He published the results of his experiments in two small books with the titles *Experiment on Plant Hybrids* (Brünn, 1866) and *On Hieracium-hybrids Obtained by Artificial Fertilization* (Brünn, 1869). Working with more than 60,000 plants, he discovered that they cross according to certain rules, the so-called hereditary laws, named after Mendel as "Mendel's laws."

Mendel had shown through these experiments that all "new" characteristics that we obtain by crossing members of the same species are only new combinations of the characteristics already evident or hidden in the ancestors. This means that crossing or breeding does not change the biological potential of parents, it only combines it to form new variations of already present potentialities. It is one of the ironies of history that in the wake of late nineteenth-century evolutionary thinking Mendel's discovery was completely ignored. When in 1900 European botanists discovered these laws for a second time and checked the literature, they found that Mendel had obtained both experimental data and developed a general theory explaining the data thirty-four years earlier.[3] Today, however, it is recognized that Mendel gave biologists the first critical experimental method for the study of heredity. Mendel's results made it plain that selective breeding could not account for the evolutionary progress Darwin had claimed.

Neither could environmental pressure, as Lamarck had once proposed, account for the biological evolution of life. For instance, when we transfer a butterfly in its chrysalis stage from a moderate environment into a cooler environment, the resulting butterfly will develop a hairy film. The same happens when we transplant a flower plant that grew in good soil and in warm temperature to poor soil and a cold environment. It will grow leaves which are smaller and covered with a hairy film. However, none of these para-variations is an actual mutation which results in a change of inherited characteristics. When we return the specimens to their traditional environment, the "new" characteristics disappear. Even human biological constitution remains unaffected by changes in the environment. As we know from the Old Testament, circumcision is a very old custom among Jewish people. However, thou-

3. See the instructive paper by Clarence P. Oliver, "Dogma and the Early Development of Genetics," in *Heritage from Mendel*, ed. Royal Alexander Brink (Madison: University of Wisconsin Press, 1967), 3f, and Futuyma, 22f.

sands of years after the first Jew was circumcised, this operation must still be performed, because the biological makeup has been unaffected by this custom. Or consider the Oriental custom of crippling the feet of women in their early childhood so that they walk with graceful small steps in tiny sandals. When the Communists banned this cruel custom in China, the next generation had feet as normal as women of all other nations. If none of these changes of the environment resulted in a change in the biological constitution, a clue to the evolutionary progress of life could perhaps be gained from a further investigation of the causes for "random variations." We remember that according to Darwin these variations are in part responsible for biological evolution. Indeed, a major breakthrough occurred when the genetic basis responsible for such "variations" was discovered.

At the Fifth International Congress of Genetics in Berlin in 1927, the American geneticist Hermann J. Muller (1890-1967) read a paper entitled "The Problem of Genic Modification" in which he reported on experiments with the fruit fly *Drosophila melanogaster* with which he had worked since his graduate studies under Thomas H. Morgan (1866-1945) at Columbia University. Subjecting the *Drosophila* to X rays, Muller gained an abnormal percentage of gene mutations, both in sperm and in eggs, and in the ovarian cells of the fly. Some of the mutations were lethal; others led to sterility. It was discovered that in some flies the linear order of some genes was rearranged, a characteristic which was inherited, and that often the X rays had a fractional effect on the genes. Muller also observed that the intensity of these effects varied with the X-ray dosage to which the objects were exposed. Now it seemed clear that the variations Darwin had postulated as the basis of the evolutionary change are actually mutations of the genetic structure of living species which in turn are handed on to subsequent generations. Darwin was also right when he claimed that such "variations" occurred at random, since it had been shown with *Drosophila* and other objects of investigation that mutations are generally nondirected, with detrimental constellations far outnumbering favorable ones.

The theory of evolution became essentially a synthesis of Darwinian thought and genetic data. The principal source of hereditary variations was found in genetic mutations and the random shuffling involved in sexual reproduction. Once Muller discovered that mutations could be artificially induced by exposing *Drosophila* to radiation, he wondered "whether the mutations that occur in untreated material are caused by radiation of a similar type from radioactive substances naturally occurring."[4] He concluded that because of the low concentration of natural radiation, most of the mutations in untreated flies cannot be caused by this source. Of course, this result was somewhat disappointing, especially since the artificially induced mutations were of the same kind as the spontaneous ones. Muller, however, concluded that "natural radioactivity, while of no consequence in flies, may appreciably influence human mutation frequency. For the long duration of the human generation sometimes allows the reception of ten or more *r* [roentgen]. Thus, under special conditions, the amount might conceivably be enough to be significant in evolution."[5]

Today we know of many more mutation-causing factors, in addition to X rays and natural radioactivity. For instance, all ionized radiation, such as ultraviolet light, cosmic radiation, and under certain circumstances even visible light, can lead to mutations. An increase in temperature and certain chemicals, such as formaldehydes and alkaloids, can also increase the rate of mutations. Due to the large number of genes in humans, it is estimated that "close to 20 per cent of all people will carry one or more newly arisen mutant genes."[6] Fortunately most of these mutations are either recessive, and thus are not noticed in the human phenotype, or have such slight effects that they are taken as normal variations in human development.

With the discovery that new inheritable characteristics are due to a change in the genetic structure of a living being, the origin of evolutionary change was withdrawn from our eyes and relocated in the molecular realm. The genetic information had been localized in the macromolecule described in chemical terms as deoxyribonucleic acid (DNA). These DNA molecules are responsible for replication and control of developmental processes. DNA contains all the genetic information of an organism and can be duplicated as often as needed.

4. Hermann J. Muller and Lewis M. Mott-Smith, "Evidence That Natural Radioactivity Is Inadequate to Explain the Frequency of 'Natural' Mutations" (1930), in *Studies in Genetics: The Selected Papers of H. J. Muller* (Bloomington: Indiana University Press, 1962), 278.

5. Hermann J. Muller, "The Role Played by Radiation Mutations in Mankind" (1941), in *Studies in Genetics*, 549f.

6. Theodosius Dobzhansky, *Mankind Evolving: Evolution of the Human Species* (New Haven: Yale University Press, 1962), 50.

The genetic code, which works according to rather simple and uniform principles, regulates the developmental and hereditary processes. But this does not result in uniformity or monotony of life. For instance, each human being has in itself forty-six chromosomes, and each single human chromosome forms a string of thousands of genes containing altogether more than 3 billion nucleotides. The DNA contained in just one human cell is approximately two meters long, and is usually wound up in this microscopically small cell and has a diameter of only 0.001 to 0.1 mm. This shows how complicated duplication of one cell is and how many possibilities there are in the hereditary process. In each successive generation the whole genetic input of two parents is available. The genes contained in a single cell, however, contain the total genetic information of a living being, and the regulated expression of this information determines the ontogenesis, the instinctive behavior, and the physiological structure of each forthcoming *individuum*. Yet the total genetic information of any cell is never available at one time and continuously effective. This is so because through certain regulative mechanisms a part of it remains inactive so that different cell types can develop, for instance a liver or skin cell.

Mutations are caused if DNA molecules are damaged or changed (gene mutation and chromosome mutation) or if chromosomes are distributed in a mistaken way. For example, if one extra chromosome appears in the genome, there will appear mongoloid features in a human being because of this trisomy (Down syndrome; gene mutation). Considering the complexity of the genetic information, we should be surprised not that mutations exist but, to the contrary, that such "mistakes" occur so rarely. Let us, for instance, modestly assume that a human parent has a set of one thousand genes, each of which is able to assume ten different forms. The offspring would be the result of a choice of 10^{1000} different combinations, which means there were more possibilities for variations than there are atoms in our universe. Small wonder that each human being has its own unique genome (genotype), its own individual face, and its own unique fingerprint. Each new being is a strictly novel occurrence.

There are different and opposing hypotheses concerning the actual process through which evolution has been occurring. Representatives of the neo-Darwinian school, Julian Huxley (1887-1975) and George G. Simpson (1902-84) to name but two, claim that evolution consists of an accumulation of small changes.[7] Others, such as Theodosius Dobzhansky (1900-1975) and Ronald A. Fisher (1890-1962), suggested that the small mutations observable in laboratory experiments are not the cause of evolution. Dobzhansky, for instance, asserts that "mutations continue to arise in man, even as they have been since the dawn of time. They are the raw materials from which natural selection gradually built the genetic endowment of the human species. Beneficial mutants are, however, a minority."[8] Scholars such as Richard B. Goldschmidt (1878-1958) and Otto H. Schindewolf (1896-1971) went one step further. They rightly reminded us that laboratory experiments have only resulted in changes within the species, but not in formations of new species. "Microevolution does not lead beyond the confines of the species."[9] Paleontologists have found only a few fossils indicating a transition between species, much less between major types. (An exception is the *Archaeopterix lithographica,* a petrified dinosaur with birdlike features found in the limestone of the Upper Jurassic, i.e., approximately 140 million years ago. The pigeon-sized skeleton resembles a transition between dinosaurs and birds with pointed teeth in the beak, claws on the wings, and a long tail spine.) Yet they assumed that "species and the higher categories originate in single macroevolutionary steps as completely new genetic systems." However, it is difficult to find much evidence for such radical mutations.

Darwin already cautioned that there are still many "missing links." The unveiling of past generations in the evolutionary chain is largely haphazard. We might add that some details will perhaps always remain

7. Cf. George G. Simpson, "The History of Life," in *Evolution after Darwin: The University of Chicago Centennial,* ed. Sol Tax, vol. 1, *The Evolution of Life: Its Origin, History, and Future* (Chicago: University of Chicago Press, 1960), 166f.; and Julian Huxley, *Evolution: The Modern Synthesis* (New York: Harper, 1943), 115.

8. Cf. Dobzhansky, 287; and Ronald A. Fisher, *Creative Aspects of Natural Law* (1950), in *The Collected Papers of R. A. Fisher,* ed. J. H. Bennett (Adelaide: University of Adelaide, 1974), 5:182.

9. Richard Goldschmidt, *The Material Basis of Evolution* (Paterson, N.J.: Pageant, 1960), 396ff., for this and the following quote; and Otto Schindewolf, *Basic Questions in Paleontology: Geological Time, Organic Evolution, and Biological Systematics,* trans. Judith Schaefer (Chicago: University of Chicago Press, 1993), 214, who opts for a "modification of the type" instead of a "modification of the species." Yet Futuyma, 24f., talks about an "evolutionary synthesis" which advocates "gradual change" in combining Darwinian evolution with genetics.

in the stage of hypothesis because the genetic processes of past aeons are forever withdrawn from our eyes. But it would be unfair to claim that evolutionary theories are the result of wishful thinking. Nonetheless, it is becoming more and more apparent that (1) there is an irreversibility of genetic steps. If a new level has been attained in the history of evolution, for instance the breathing through lungs in land animals, then a subsequent retreat into the sea, for instance of dolphins, does not result in a return from lungs to gills. The breathing through lungs remains even if, as in the case of dolphins, it does not yield an evolutionary advantage. (2) One has realized that there is a feedback of information from the outside onto the genetic makeup. For instance, bacteria continuously become resistant to antibiotics, a fact which can be traced through a change of their genetic code as a result of them having been exposed to antibiotics. The genes are undoubtedly the main carriers of the evolutionary processes. Yet they are not totally independent from the environment in which the genetic information is expressed.

b. Unity and Evolvement of the Living Species

The various life-forms show such a fundamental and astounding unity that this suggests both a relationship of all living beings and a common evolution. The first set of evidence for a unity of life-forms can be derived from comparative anatomy. With the exception of a few parasitic forms, the roughly one-quarter million different species of flowering plants have the same basic root structure, stem-bearing branches, leaves containing chlorophyll, and flowers made up of modified leaves. They also live in the same way. They absorb dissolved chemicals through their roots and integrate them with the help of sunshine, chlorophyll, and carbon dioxide into their own substance. The amazing variety of flowers we encounter seems to go back to one basic structure or to one basic ancestor. When we survey the roughly 800,000 species of insects, we again detect a basic unity in their structural design. The design of their bodies all shows the same division into head, trunk, and abdomen. They have three pairs of legs and two pairs of wings and their mouth parts are again built alike, although some use their mouths for sucking and others for biting.

Coming to vertebrates, we notice yet again a surprising similarity. Though the bones have modified in accordance with their different uses, the fundamental structural plan is the same. Whether lizard, mole, eagle, wolf, ox, or gorilla, they all have similar skeletons, forming one big family. Whether with four feet, two feet and two arms, or stunted feet, they all look somehow similar, and the conclusion can hardly be avoided that there is a basic unity in this individual diversity. Often the same parts, however, are only present in either further developed or reduced varieties. In some fish, for instance, the electrical potential in specialized muscles has been increased to such an extent that through electric discharge other animals can be killed or paralyzed. Quite often the parts have receded to so-called rudimentary parts. When we look at a whale, for instance, its hind legs are hardly noticeable. Similarly, the wings of an ostrich are no longer fit for flying. Humans themselves have more than thirty of these rudimentary parts, the most bothersome being the appendix. For us it no longer serves a useful function, but in plant-eating mammals it is a sack in which bacteria digest the cellulose cell walls of vegetable food. One would not be far from the truth in assuming that these rudimentary parts point to a common ancestor from which different species developed in different directions.

A second type of evidence for a basic unity of living species can be derived with the help of biochemistry. The most striking evidence here can be obtained from serology.[10] Blood transfusions show that the blood from one human being can be received by another without complications for the latter if both belong to the same blood group. However, if a person were to receive the blood of a dog, signs of being poisoned would immediately be evident. The body would fight the foreign blood by producing antibodies or antitoxins which react against it and the morbific agents it contains. The antibodies immunize a body against foreign blood and are constantly present in the blood after such an immunization. In a separator we can sort the red blood corpuscles from the pale yellow liquid of the blood serum that contains the antibodies. If we take the blood serum which contains the antibodies and treat it further with blood from a dog, the animal species against which the blood was immunized, the serum thickens and the foreign blood plasma falls out as

10. Cf. Philip L. Carpenter, *Immunology and Serology* (Philadelphia: W. B. Saunders, 1956), 39, who cites interesting evidence. For an extensive discussion of this procedure, its results and limitations, cf. R. D. Martin, *Primate Origins and Evolution: A Phylogenetic Reconstruction* (London: Chapman and Hall, 1990), 587-602.

an insoluble product. Obviously the antibodies react against the type of blood by which they were caused. Similarly a vaccination helps against a specific disease against which one was vaccinated. But we would not expect, for instance, that a polio vaccination helps against measles. Yet it was discovered that human serum or any other serum which was treated with dog blood will also react, though more weakly, with wolf blood, and still more weakly with blood of foxes or hyenas. This means that the blood serum also reacts with blood that is closely related to the blood with which it was treated. The stronger the reaction, the closer is the chemical affinity. When we label the strongest reaction with 100 and the weakest with 0, we get the following scale: human, 100; gorilla, 64; orangutan, 42; baboon, 42-29; and spider monkey, 29. We notice that there is a closer chemical affinity between human blood and the blood of anthropoid apes than between human blood and the blood of other species. Of course, the scale does not prove that there exists an actual blood relationship, but such a conclusion would not be unfounded.

We should mention two other striking phenomena, connected with the enzyme cytochrome c and the pancreatic enzyme insulin. Cytochrome c molecules are albumins with an iron-containing *hem* group, similar to the hemoglobin, responsible for the red color of the blood. Cytochrome is essential for the oxygen transfer in the breathing process within cells. It has been discovered that the albumin chains of cytochrome c are more similar to each other the closer organisms and their functions are related to each other. "Cytochrome c . . . consists of a polypeptide chain of 112 amino acids. The first 8 occur only in bacteria. . . . Humans differ from rhesus monkeys by only one amino acid in position 66 corresponding to a difference of 0,9%, and from horses by 11 amino acids, corresponding to a difference of 10,6%. They differ from chickens by 22 amino acids (21%), from rattlesnakes by 29 amino acids (32%), and from beer yeast by 62 amino acids (59%)."[11]

We do not need much fantasy to assume a common origin of the cytochrome c from which those different variations of this enzyme developed. A similar phenomenon occurs with the enzyme insulin, which is made up of fifty-one amino acids. Again the sequences (homology) of the amino acids vary with the different species we investigate. A closer

11. According to the textbook by Cesare Emiliani, *Planet Earth: Cosmology, Geology, and the Evolution of Life and Environment* (Cambridge: University Press, 1992), 377 fig. 19.4.

similarity of the sequences could be interpreted as a closer affinity be-tween two species, and a not-so-close similarity as a more distant rela-tionship between the two. Humanity has always taken notice of the af-finity of all living beings. From antiquity to the most recent times, many philosophers and theologians postulated a great chain of being which ties together everything that was created, from stones, plants, and ani-mals to the heavenly beings. Humanity was assigned a mediating posi-tion between animals and angels. Through the evolutionary theory this unity of all living beings has again been emphasized. Yet now it was not only considered to be a unity of everything created, so to speak, on one plane, but temporality was included to show that everything had a com-mon origin in time.

Paleontology tries to discover this temporal connection between various living beings which have existed at various times by investigating the traces of living beings who have long been extinct. Decisive for this endeavor is an exact dating of the individual finding. This is possible in a rather general way through the classification of secondary sediments, for example, limestone or sandstone formations. A more exact dating results from measurements of radioactive decay with the findings, for instance, through the isotope carbon 14, uranium 234, and beryllium 10. In various sediments fossils are enclosed of petrified plants or animals which are long extinct. The younger the sediments, the more fossils we find and the more they resemble species which still exist today. The oldest fossil-containing layers are in the Precambrian period and date back about 3.5 billion years. These objects seem to be fossilized bacteria and blue-green algae.[12] One billion years ago, it is safe to say, we find green algae and fungi. The first trace of an animal, a worm, is found 800 million years ago in the Precambrian period. In the Cambrian period, 570 million years ago, nonvertebrates are already well differentiated. In the next layers, Ordovician and Silurian, fishes are the highest animals. Amphibians ap-pear in the Devonian period and reptiles in the carboniferous strata that also contain our coal. Finally mammals are present in the Cretaceous pe-riod (barely 130 million years ago). Humans themselves are latecomers,

---

12. For an extensive discussion of the oldest fossils, cf. M. G. Rutten, *The Origin of Life by Natural Causes* (Amsterdam: Elsevier, 1971), esp. 219ff. Cf. also Raymond Enay, *Palaeontology of Invertebrates*, trans. Thomas Reimer, foreword by Jean Aubouin (New York: Springer, 1990), 14f.

and we must confine ourselves to the last 20 million years to find traces of our first ancestors.

It should be a matter of course that we can no longer interpret fossils in the same manner as the British naturalist Philip H. Gosse (1810-88). Two years prior to the publication of Darwin's *The Origin of Species,* Gosse attempted in his book *Omphalos* to connect geological findings which would support an evolutionary tendency with a literal interpretation of a creation within six days. The findings of geology proved for him only that God had created the world "with fossil skeletons in its crust — skeletons of animals that never really existed."[13] We can no longer deny the existence of prehistoric living beings. Bone analysis and the arrangement and structure of teeth of fossil findings allow us today even to estimate the climate in which these fossilized animals lived and the diets they enjoyed. The kind of climate and food usually corresponds with the environment, which we can reconstruct from the sediments in which the findings are discovered.

Even the catastrophe theory of the French naturalist Baron Georges de Cuvier (1769-1832) is no longer tenable for us. He claimed in his book *Essay on the Theory of the Earth* (1818) that sudden floods, of which Noah's flood was the most recent, had destroyed entire species of organisms. The reason for introducing catastrophes into the evolutionary process, he contended, was because each species was so well coordinated that it could not undergo developmental change. Therefore once a catastrophe had wiped out an entire population of all the countries presently inhabited, the devastated land would be repopulated by a small number of animals, if not by humans "that escaped from the effects of that great revolution."[14] Though disallowing a succession of different species on a global scale, Cuvier's theory at least attempted to account for the succession of different species in limited areas.

Cuvier, however, was not completely wrong when he assumed a fixed rigidity of species. While today fossils of extinct species are usually explained by means of evolutionary change, it is also assumed that

13. Philip H. Gosse, *Omphalos: An Attempt to Untie the Geological Knot* (London: John van Voorst, 1857), 347f.

14. Georges Cuvier, *Essay on the Theory of the Earth*, with *Mineralogical Notes* by Robert Jameson and *Observations on the Geology of North America* by Samuel L. Mitchill (New York: Kirk & Mercein, 1818), 166. It is not without interest that the editors used as the frontispiece a drawing of the *Archaeopterix lithographica.*

they were no longer strong enough to survive and consequently were displaced by more advanced and better-equipped offspring of the same species. Sometimes even whole species died out and were supplanted by other, presumably better equipped species, as we see in the disappearance of the dinosaurs. Only places which early in their history were cut off geographically from the main evolutionary struggle for existence, such as the Galápagos Islands or Australia, could retain some archaic relics and go their own way within the overall evolutionary process. Thus Marsupialia, such as the kangaroo and the koala, could evolve in abundance in Australia, and the Galápagos Islands are still inhabited by huge turtles that are long extinct in other, less secluded areas. The conclusion is almost inescapable that there is an evolutionary movement within a species that proceeds from simple to more and more complex. This upward movement is traceable by evidence of fossils both within the individual species and from one species to another.

There also seems to be an evolution from mammals to human beings, the youngest traceable living species. The findings of archaeology and paleoanthropology seem to indicate that there is an upward-directed development within human ancestry. First is the split of the hominoids, meaning those who resemble human beings, into pongids, those similar to monkeys, and hominids, those similar to human beings, approximately 24-18 million years ago. Then there is the emergence of the prehuman *Ramapithecus*, documented in India and East Africa; then *Australopithecines* approximately 2.5 million years ago, who show human potential and are from South Africa and East Africa as well as Indonesia; and finally *Homo sapiens*, who emerged approximately 250,000 years ago and has its most famous representative in the so-called Neanderthal man. Contrary to occasional claims of an African genesis or of the origin of humanity in China, we should perhaps follow the German anthropologist Gerhard Heberer (1901-73), who assumed an "animal-man transition field."[15] This phase of hominization lasted approximately

---

15. Cf. Gerhard Heberer, "Die Herkunft der Menschheit," in *Propyläen Weltgeschichte*, ed. Golo Mann and Alfred Heuß, vol. 1 (Berlin: Propyläen, 1961), 127ff. Heberer talks here of an "animal-man transition field" in which the tool-user became a toolmaker. For a good introduction to the evolution of early humans and the tentative character of any conclusions, cf. B. A. Wood, "Evolution of Australopithecines," 231-41, and C. B. Stringer, "Evolution of Early Humans," 241-51, both in *The Cambridge Encyclopedia of Human Evolution* (1992).

600,000 generations and signifies the transition from humanlike beings *(Ramapithecus)* to humanity itself *(Australopithecus)*. This would also mark the time span in which anthropoid apes emerged. The anthropoid apes we have today, such as gorillas, chimpanzees, and orangutan, are therefore not part of our human prehistory. Among *Australopithecus* and even more among *Homo sapiens* we notice an astounding variety. We are coming ever closer to our immediate ancestors and can no longer talk precisely about different "species." They are "evolutionary processes within a single species."[16]

If one nation prevails over another, usually the gene pool of the losing nation is not eliminated but merges with the winning nation. Similarly, we may understand the various groups which sometimes merge with each other. A good example is the relationship of *Australopithecus* to *Homo sapiens*. After many battles and amalgamations our modern variety of humanity gradually emerged. Essentially this is only distinguished by three races, the Caucasoids, the Mongoloids, and the Negroids, whereas in very distant areas at the fringes of continuous settlements there still exists an earlier kind of humanity, such as the Ainu in northern Japan, the Papuans in New Guinea, and the Aborigines in Australia. After we have attempted to trace the evolution of living beings up to humanity, we then must ask how we can understand the transition from the nonliving to the living nature from a scientific perspective. Did everything there occur in a "natural" way? Or do we have to reckon with otherworldly powers?

c. The Origin of Life

A Latin proverb says: *Ex ovo ovum* (an egg comes always from an egg), meaning that life always presupposes life. Throughout the Middle Ages, however, it was commonly accepted that many lower animals, such as rats, worms, bees, fish, and even mice, were spontaneously generated from waste products. Once one discovered, for instance, that the eggs of flies are the connecting band between one generation and the next, this Latin proverb reacquired increased significance. Around the middle of the nineteenth century it was common opinion that the life and prod-

16. Dobzhansky, 191.

ucts of living organisms can only be generated from living organisms. The conviction of a fundamental difference between the organic and the inorganic, i.e., between living beings and substances that are derived from dead matter, was severely shattered in 1828 by the work of the chemist Friedrich Wöhler (1800-1882). He succeeded in producing urea from ammonium cyanate. In 1807 the Swedish chemist Jöns Jakob Berzelius (1779-1848) had distinguished between two classes of chemical substances, the organic and the inorganic, and claimed that inorganic substances occurred in nature independently of life while organic substances are produced only by living beings. Wöhler's results refuted the idea that organic substances could be produced only from organic beings through the presence of a "vital force." Ammonium cyanide was considered an inorganic chemical, while urea was definitely an organic substance. Yet the ardent defenders of vitalism, claiming a life force necessary to bring forth organic substances, did not concede defeat through the results of Wöhler's experiments. They now claimed that the urea, which is also present in a natural way in urine, is only an excretory substance. It is the result of a breakdown and not of a synthesis that leads to life. Soon thereafter, however, other chemists succeeded in synthesizing organic substances from inorganic ones, and today many known organic compounds can be gained this way. Yet it is still customary to distinguish between inorganic and organic substances, whereby the latter ones are defined as compounds containing carbon atoms in their molecules. Yet the vitalistic claim of a strict distinction between the two has been abandoned.

We still might assert that there is a fundamental difference between a dead organic substance and a living *individuum*. But again it is difficult to draw a strict line between the two. It is even difficult to arrive at definite characteristics which distinguish a living *individuum* from inanimate matter. For instance, reproduction, growth, irritability, and self-regulation are not exclusive features of life. Scientists do not only talk about growing animals and bacteria, they also talk about growing crystals in a chemical solution. Once the crystallization process has started, a crystal grows according to a certain pattern determined by the chemical structure of the crystal. If we disturb a crystal in its growth, perhaps through the presence of a solvent or a corroding agent, it develops in an abnormal way, showing a depression on the face of the crystal.

Viruses may serve as another example to illustrate how difficult it is to distinguish the living from the dead. Viruses, such as the tobacco mosaic virus, can be crystallized like most organic and inorganic substances. Once crystallized they show no observable traces of life. They consist essentially of nucleic acids and proteins arranged for each type of virus in a characteristic form. While most of the polypeptidic chains of protein molecules have a molecule weight of between 12,000 and 35,000 dalton (i.e., they consist of approximately 2,000 atoms), the smallest viruses, such as the tobacco mosaic virus, have a molecular weight of approximately 40 million dalton (more than 1 million atoms). This huge virus "molecule," however, is a living entity and not dead matter. If put on the right host material, it reproduces, and within 30 minutes can "give birth" to another 200 newly formed viruses. It does not matter how long it was in its crystallized stage or in an environment in which it could not reproduce. Scientists have found viruses, caught thousands of years ago in glacial ice, which, when put into a favorable environment, were as reproductive as others. Viruses are adjustable to their environment and can also undergo mutations. This has been painfully observed when a newly developed medicine is suddenly no longer effective against a certain virus. The virus has become immune or has even undergone a mutation and thus has escaped the deadly threat of the medicine. The influenza viruses that were first isolated in 1933 seem to undergo continual changes, always calling for new vaccines or immunizations against their attacks.

Of course, there is still a big step from viruses to amoebas or other one-cell beings. Yet we have noticed that the distinction between the living and the nonliving is in flux and can even be bridged in laboratory experiments. This could indicate that life originated from nonliving matter. Experiments in this direction have yielded astounding results. For instance, the American physical chemists Harold C. Urey (1893-1981) and Stanley L. Miller (b. 1930) reproduced this primeval atmosphere in their laboratory in Chicago in 1954.[17] While our present atmosphere consists largely of oxygen and nitrogen, this primeval atmosphere at the beginning of our earth was totally free of nitrogen and

17. Stanley L. Miller, "Production of Some Organic Compounds under Possible Primitive Earth Conditions" (1955), in *Synthesis of Life*, ed. Charles C. Price (Stroudsburg, Pa.: Dowden, Huchinson & Ross, 1974), 7-17.

consisted largely of hydrogen and hydrogen compounds such as methane, ammonium, and simple water vapor. For one week they transmitted sparks through a capsule containing a mixture of these gases and subjected the capsule to intense ultraviolet radiation, thus duplicating events which could have been expected in the primeval atmosphere in which intense thunderstorms and high sun radiation were likely common. To the surprise of all, they produced organic compounds, including some of the simpler amino acids, in these experiments. This result is even more remarkable when we consider that amino acids are the building blocks out of which proteins are composed, and proteins are common to all forms of contemporary life.

Though it is a long way from simple amino acids to a living cell or even to human beings, a "natural" transition from inanimate matter to living beings might now, in principle, be conceivable. The argument for a natural generation of life gained new strength when the American biochemist Sidney W. Fox (1912-98) reported in 1960 that under presumably primeval conditions on earth spontaneous production of proteinoids is possible.[18] He claimed that these proteinoids differ only slightly from the natural proteins of low molecular weight. The molecular weight depended on the temperature at the time of origin, allowing for molecular weights up to 8,600 dalton compared with a value of 6,000 dalton for insulin. The proteinoids also have a tendency to assume cell-like shapes in aqueous solutions, and their amino acid units showed some degree of order, which is the aim in repeated polymerizations. Fox, however, was very cautious in his conclusions. He was not convinced that at present we can experimentally demonstrate how life began. Yet if it were possible to start life by producing a cell which metabolizes and reproduces itself, he argued, we might be in a better position to determine whether the conditions under which this cell was generated are "conditions associated with the current earth, with what we believe to have been the prebiological earth, and with conditions prevailing on other planets."[19] But he continued: "Although we can with certainty say only that life arose at least once, there is increasing reason to believe that life can, or even must, arise in many places at many

18. For the following see Sidney W. Fox, "How Did Life Begin?" *Science* 132 (July 22, 1960): 200-208, here 206.
19. Fox, 207, for this and the following quote.

times." Of course, such a hypothesis of a polygenetic origin of life always presupposes a primeval atmosphere.

It has been shown that as soon as (laboratory) conditions become oxidizing (i.e., a sizable amount of free oxygen is present), such as in our present earth atmosphere, no organic compounds, and this means no amino acids, are formed.[20] This seems to suggest reducing conditions for the prebiotic synthesis of amino acids. If amino acids are already available, however, then proteinoids are frequently produced in an atmosphere in which oxygen is present. The discovery of archaebacteria seems to affirm this hypothesis about the origin of life. Archaebacteria do not have a nucleus and therefore chronologically come before eukaryotes in the origin of life since they have already a nucleus. These three — archaebacteria, eukaryotes, and bacteria — represent all life known to us. Archaebacteria can live in very hot temperatures (exceeding 75 degrees Celsius) and often thrive where there is no oxygen, for instance, in a solution which contains a preponderance of sulfur or methane compounds. All this points in the direction of the so-called primeval atmosphere.[21]

After the essential building blocks of life are traced in the so-called primeval atmosphere of our earth in laboratory experiments, the question needs to be solved as to how these prebiotic components became self-duplicating, changeable molecular systems which interact with their environment. Here the biggest difficulty is explaining the spontaneous origin and development of the genetic code. The French Nobel Prize winner for medicine, Jacques Monod (1910-76), summed up the issue of the origin of DNA or of the genetic code very aptly: one possibility is that there is stereochemical affinity between a certain unit of the genetic code and a specific amino acid. If this were true, there would be a necessity for the structure of the genetic code as it is. Another possibility is that "the code's structure is chemically arbitrary: the code as we know it today is the result of a series of random choices which grad-

---

20. Cf. Carl Sagan, "The Origin of Life in a Cosmic Context," in *Cosmochemical Evolution and the Origins of Life. Proceedings of the Fourth International Conference on the Origin of Life and the First Meeting of the International Society for the Study of the Origin of Life. Barcelona, June 25-28, 1973*, vol. 1, *Invited Papers*, ed. J. Oré, S. L. Miller, C. Ponnamperuma, and R. S. Young (Dordrecht: D. Reidel, 1974), 498.

21. Cf. Karl O. Stetter, "Manche mögen's heiß: Mikrobielles Leben an der obersten Temperaturgrenze," *Blick in die Wissenschaft: Forschungsmagazin der Universität Regensburg* 2, no. 3 (1993): 14. Cf. also Futuyma, 169.

ually enriched it."[22] Francis Crick (b. 1916) seems to opt for the latter possibility when he suggests that there were three steps in the formation of the genetic code:

1. The Primitive Code, in which a small number of amino acids were coded by a small number of triplets in the DNA structure.
2. The Intermediate Code, in which these primitive amino acids took over most of the triplets of the code in order to reduce nonsense triplets to a minimum.
3. The Final Code as we have it today.

This means that the number of twenty amino acids and the actual amino acids specified through the code are, in part at least, due to historical accident. Yet Crick does not want to leave everything to chance. At first, he declares, the Primitive Code would code specifically for only a few amino acids. However, as the process of development of the genetic code proceeded, "more and more proteins would be coded and their design would become more sophisticated until eventually one would reach a point where no new amino acid would be introduced without disrupting too many proteins. At this stage the code would be frozen," since further development would no longer give any advantage.[23] This would mean that the genetic code evolved, from some stage onward, according to the interaction with the enduring proteins it specified. These conclusions are not really surprising, since the building blocks of the genetic code, namely, amino acids, nucleotide bases, and sugars, are formed with surprising ease from a few single molecules such as formaldehyde, hydrogen cyanide, and cyanoacetylene.[24] Perhaps biochemist Leslie Orgel (b. 1927) is right when he concludes that neither proteins without nucleic acids nor nucleic acids without proteins could have developed for long without their mutual assistance. Today this seems to be the most common opinion.

22. Jacques Monod, *Chance and Necessity: An Essay on the Natural Philosophy of Modern Biology*, trans. Austryn Wainhouse (New York: Alfred A. Knopf, 1971), 143.

23. See Francis H. C. Crick, "The Origin of the Genetic Code," *Journal of Molecular Biology* 38 (December 28, 1968): 375, who also describes very well his thesis concerning the development of the genetic code.

24. Cf. Leslie E. Orgel, "Evolution of the Genetic Apparatus," *Journal of Molecular Biology* 38 (December 28, 1968): 381f.

We must remember, however, that there is an immense difference between protein and proteinoid, the latter being conceptually a pre-protein, and that by contact with water proteinoids form cell-like micro-structures (proteinoid microspheres) in vast numbers. Sidney Fox and others have shown that these microspheres produce budlike appendages which, when separated from the microspheres, behave "as physical nuclei around which polymers accrete to yield full-size acidic proteinoid microspheres."[25] These newly emerged microspheres in turn frequently form junctions with one another. Tiny pockets of proteinoids, called endoparticles, separate within the walls of the microspheres and pass through these communicative junctions from one microsphere to another. When released from their acidic microspheres to the outside through holes in the walls, the endoparticles function as accretion nuclei which accumulate fresh acidic proteinoid of their own kind to form a "second" generation of microspheres. Since the microspheres are still prior to the actual protein and yet have cell-like features, it seems likely that between proteinoids and later protein some kind of protocell emerged. This protocell was then able to make the kind of inter-nucleotide bond (communicating and specifying channel) that characterizes nucleic acids. Once we arrive at the level of proteins, however, they force corresponding nucleic acids to regulate their own structures.

The avenue along which the genetic code evolved now seems even more clarified by observing a mechanism by which a protein directly controls expression of its own structural gene. "The essence of this regulatory mechanism is that a protein specified by a given structural gene is itself a regulatory element which modulates expression of that very gene. Thus, the protein regulates the rate at which additional copies of that same protein are synthesized as well as the rate of synthesis of any other protein encoded in the same operon."[26]

Though growth and development of protein chains seem to stem from these choices themselves, once growth and development have taken place, a correlated genetic code insures that the newly acquired features are preserved for further use. The nucleic acids seem to func-

25. Laura L. Hsu and Sidney W. Fox, "Interactions between Diverse Proteinoids and Microspheres in Simulations of Primordial Evolution," in *BioSystems* 8 (1976), esp. 8:90.

26. Robert F. Goldberger, "Autogenous Regulation of Gene Expression," *Science* 183 (March 1, 1974): 810.

tion like superadministrative macromolecules that insure that what is manifested is selected once it survives.

How the nucleic acids and the "corresponding" proteins worked together to the actual breakthrough from the prebiotic stage to the biotic is still a matter of conjecture. Sidney W. Fox and Klaus Dose (b. 1928), for instance, claim:

> Two approaches have been evident in the modeling of the origin of nucleic acids. One is the search for models of pre-biotic nucleic acid. The other is the conceptualization that many biologists have entertained — the concept that nucleic acids as superadministrative macromolecules came after protein and cell. This kind of thinking is Darwinian in the premise that what survives is selected after it is manifested, rather than that it is directed before it is expressed. Nor is this mode of thinking inconsistent with phenotypic expressions in succeeding generations. The range of possibilities can be determined by protein biosynthesis, and selected by the nucleic acids.[27]

Yet we wonder whether these two approaches are that far apart, especially if cell is understood as "primitive cellular constellations" and proteins as "proteinoids."

Most probably many different occurrences independent of one another were needed to give rise to life on our primitive earth. Yet eventually only one line survived and developed, and every organism on earth today has descended from this line. This hypothesis can be supported with a very interesting phenomenon. Chemists distinguish between levorotatory (L) and dextrorotatory (D) organic substances. These chemicals are identical except that they are built mirror-like, such as a left-hand glove is different from a right-hand glove. When these optically isomeric molecules are produced in the laboratory from combinations that do not show isomeric properties, half the resulting molecules are levorotatory and half dextrorotatory. However, in nature there is no symmetry, because one form usually dominates the other. For instance, tartaric acid produced in a laboratory consists of 50 percent levorotatory and 50 percent dextrorotatory tartaric acid molecules. In natural fruit

27. Sidney W. Fox and Klaus Dose, *Molecular Evolution and the Origin of Life*, foreword by A. Oparin (San Francisco: W. H. Freeman, 1972), 253. This passage is omitted in the 1977 revised edition.

juice, however, we find only dextrorotatory tartaric acid molecules. Similarly our whole biosphere is "constructed apparently from *only* the L-amino-acids and D-sugars," though "a living world of left- and right-handed molecules would be equally probable."[28] All amino acids that occur in proteins have levorotatory configuration with the exception of a few small proteins (peptides) that occur in bacteria. D-alanine and some other D amino acids have been found in bacteria. Yet these peptides are toxic to other bacteria and are used in medicine as antibiotics. On the basis of laboratory experiments we might expect an equilibrium between D-rotatory and L-rotatory amino acids in living matter. Yet the discovery that the overwhelming majority of life on earth is made up of L-rotatory amino acids as the building blocks for proteins may be an important argument in favor of the unity and uniqueness of life on earth.

But we must also account for the discovery of organic matter in carbonaceous chondrites (i.e., stony meteorites consisting of claylike hydrous silicate minerals, carbonate and sulfate minerals, iron oxides, and sulfur). Though there is always the chance that these meteorites were contaminated by earth products when they fell to earth, "there is little doubt that the bulk of the organic matter in meteorites is indigenous."[29] Yet studies of meteorites show that the frequency of the L-configuration of amino acids found in them resembles that of amino acids found in recent sediments and soils on earth. This would indicate that a possible evolution of L-rotatory compounds is not confined to the earth. It has been suggested that the organic compounds in meteorites have been formed by catalytic reactions of CO, H_2, and NH_3 in the solar nebula, and that these reactions may be the source of prebiotic carbon compounds on the inner planets and of prebiotic interstellar molecules.[30] The Russian biochemist Aleksandr Oparin (1894-1980), one of the pioneers in the quest for the origin of life, even claims that the evolution of organic matter began "before the formation of the Earth — on cosmic objects such as planetesimals and particles of gas and dust. After the Earth had formed, and its lithosphere, atmosphere,

28. W. Thiemann and W. Darge, "Experimental Attempts for the Study of the Origin of Optical Activity on Earth," in *Cosmochemical Evolution and the Origins of Life,* 1:263-83, esp. 264f.

29. Edward Anders, Ryoichi Hayatsu, and Martin H. Studier, "Organic Compounds in Meteorites," *Science* 182 (November 23, 1973): 783.

30. Cf. Anders, Hayatsu, and Studier, 789.

and hydrosphere had developed, monomeric and polymeric matter became more complex. Then the first forms of life evolved, and the elaboration of their structures and metabolism continued."[31] Though such cosmic synthesis of prebiotic compounds is possible, it still does not enlighten the predominance of the L-configurations.

Yet we should also consider the discoveries connected with the Murchison meteorite, a large meteorite that fell in 1969 near Murchison, Australia. Through careful and elaborate experiments, Keith Kvenvolden and others found that "the D and L enantiomers of amino-acids in the Murchison meteorite are almost equally abundant."[32] Furthermore, Kvenvolden suggests a nonearth nature for the amino acids and hydrocarbons in this meteorite. This claim is underlined by the discovery in the meteorite of two amino acids generally not found in biological systems. These findings seem to reassure one observation made so far: organic compounds are generated with surprising ease and in an amazing manifoldness.

Yet the multitude of prebiotic substances and forms seems to diminish remarkably once we turn to the kind of life that is traceable on earth. Jacques Monod, for instance, who is convinced of a strictly mechanistic evolution of life, concedes concerning the possibility of the appearance of life on earth: "The present structure of the biosphere far from excludes the possibility that the decisive event occurred *only once*. Which would mean that its *a priori* probability was virtually zero."[33] But then he asserts that "through the very universality of its structures, starting with the code, the biosphere looks like the product of a unique event. It is possible of course that its uniform character was arrived at by elimination through selection of many other attempts or variants. But nothing compels this interpretation." Monod prefers to leave the issue open by admitting that "at the present time we have no legitimate grounds for either asserting or denying that life got off to but a single start on earth, and that, as a consequence, before it appeared its chances of occurring were next to nil."

31. Aleksandr Oparin, foreword to *Molecular Evolution and the Origin of Life*, by Fox and Dose, vii (rev. ed., v).

32. Keith Kvenvolden, James Lawless, Katherine Pering, et al., "Evidence for Extraterrestrial Amino-Acids and Hydrocarbons in the Murchison Meteorite," *Nature* 228 (December 5, 1970): 924.

33. For this and the following quotes, see Monod, 144f.

Other scientists are less hesitant to assume the unique character of a spontaneous generation of life on earth. The German physicist Pascual Jordan (1902-80), for instance, argues from the predominance of the L-rotatory substances for a monogenetic origin of life.[34] Like others, he assumes that thunderstorms and radiation by the sun caused the formation of amino acids in the primeval atmosphere of our earth. These amino acids grouped together, but still these bricks of life did not have the capability to reproduce. Then, however, he claims, a quantum jump occurred. This quantum jump caused a mutation. And in this mutation one single levorotatory molecule acquired the ability to reproduce. Thus life began, and all organic substances are descendants of this single levorotatory, reproductive molecule. Had this mutation from the unreproductive to the reproductive molecule happened at many places or in a large number of molecules, it would be much more likely that dextrorotatory and levorotatory molecules would have originated in the same amount. A mutation from which life ensues must be so unlikely that it could only happen once on our earth. Thus we have almost exclusively levorotatory organic substances in nature.

The Nobel laureate Ilya Prigogine (b. 1917) attempted a different explanation for why amino acids appear in living beings exclusively in their levorotatory form.[35] He considers it possible that the preponderance of levorotatory forms of life is due to a "neutral current" which has been discovered in elementary particle research. The strength of this current would normally be imperceptible since it is extremely weak.

---

34. Cf. for the following, Pascual Jordan, *Der Naturwissenschaftler vor der religiösen Frage,* 2nd ed. (Oldenburg: Gerhard Stalling, 1964), 337ff. Of course, this theory seems to be very unlikely, since it would exclude the existence of bacteria with dextrorotatory proteins. It is more likely that very early in the evolutionary process organisms appeared containing D-rotatory amino acids and others with L-rotatory amino acids. Some may have even contained both. Yet gradually organisms containing L-substances survived "somehow" and outpaced all the other organisms with D-configuration. Cf. George Wald, "The Origin of Optical Activity" (1957), in *Synthesis of Life,* 242, who advances this theory. He too can give no reasons why L-configurations should be preferred. Mark Ridley, *Evolution,* 2nd ed. (Cambridge, Mass.: Blackwell Science, 1996), 545, admits too: "The long time from the origin of life and of prokaryotic cells [without nucleus] to the origin of eukaryotic cells [with nucleus] — about 50% of biological history — suggests that it was an improbable, or evolutionary difficult, transition."

35. According to Harald v. Sprockhoff, *Naturwissenschaft und christlicher Glaube — Ein Widerspruch?* (Darmstadt: Wissenschaftliche Buchgesellschaft, 1992), 37f.

According to theoretical calculations, the levorotatory and dextro-rotatory forms are distinguished by the speed which determines their chemical reactions, and this is in the magnitude of 10^{-17}. This extremely low difference is supposed to explain the preferred existence of levo-rotatory configurations. One starts with the assumption that levorota-tory amino acid molecules succeeded in the prebiotic evolutionary phase because of their greater speed in the catalyzing processes. Though there is no stringent proof yet that this one-sided preference of the levorotatory amino acids in the evolutionary process actually has its cause in that difference, once again a "natural explanation" has been in-troduced for the dominance of the levorotatory amino acids and any ac-cident is excluded.

Whether we follow Monod, who leaves the issue of polygenetic ori-gin of life on earth undecided, or side with Jordan, who opts for a monogenetic origin of life on earth, both options are very much alike in their assessment of life beyond this earth. The existence of organic or even intelligent life in other parts of the universe cannot be denied a priori. As we have seen with meteorites, there are prebiotic organic compounds to be found outside our earth. Exobiology, the biology that concerns itself with extraterrestrial life, is a serious scientific enterprise and not a part of science fiction. What has been found so far, however, are building blocks leading toward life and not remains of (once) living beings. Even the bonanza of organic molecules on the surface of the moon, expected by many scientists, ended in disappointment. Scientists examining the lunar findings of the Apollo program discovered only a few amino acids, perhaps deposited on the moon by the solar wind.

Organic life originates and develops through mutations. The more mutative steps that are required to yield higher forms of life, the more unlikely parallel developments become between life on earth and possi-ble life on other planets. Thus the possibility for a similar development of life on different planets even with identical environmental conditions grows ever slimmer with each mutation. The myriad of other human-like races, of which Giordano Bruno dreamed centuries ago and which Immanuel Kant accepted as a reality, seem to be untenable specula-tions.[36] Though in our position as *living* beings we might be "rivaled" by

---

36. Cf. Antoinette Mann Paterson, *The Infinite World of Giordano Bruno* (Springfield, Ill.: Charles C. Thomas, 1970), 20, who refers to the "infinity of habitable worlds," an

other living beings beyond our earth, in all likelihood our situation as *human* beings is as unique with regard to extraterrestrial life as it is with regard to life on earth.[37] Perhaps the American anthropologist Loren Eiseley (1907-77) is right when he says about us earthlings: "Nowhere in all space or on a thousand worlds will there be men to share our loneliness. There may be wisdom; there may be power; somewhere across space great instruments, handled by strange, manipulative organs, may stare vainly at our floating cloud wrack, their owners yearning as we yearn. Nevertheless, in the nature of life and in the principles of evolution we have had our answer. Of men elsewhere, and beyond, there will be none forever."[38] We may be the crown of creation and regard ourselves as such. Yet we have nobody with whom we can talk about creation and whom we can ask for counsel on how to treat it. The responsibility for us and the environment which sustains us is solely ours.

---

idea which would have "horrified" Kepler; and cf. Immanuel Kant, *Universal Natural History and Theory of the Heavens,* introduction by Milton K. Munitz (Ann Arbor: University of Michigan Press, 1969), 165ff.

37. Cf. also William G. Pollard, *Transcendence and Providence: Reflections of a Physicist and a Priest* (Edinburgh: Scottish Academic Press, 1987), 179, who calls it an "illusion that there is anything inevitable about an evolutionary process elsewhere in the universe."

38. Loren Eiseley, *The Immense Journey* (New York: Random House, 1957), 162.

3. A Relativistic Understanding of the World

In the nineteenth century Friedrich Engels could still justifiably assert: "We have the certainty that matter remains eternally the same in all its transformations, that none of its attributes can ever be lost."[1] But such convictions are long past. We know that everything which has had a beginning will also have an end. At the end of the world there is a burned-out wreck, the lights are extinguished and all life has ended. The second law of thermodynamics, or the law of entropy, reminds us very pointedly that the different levels of energy will move ever closer to each other until the nonconvertibility of energy has attained a maximum. Within this discomforting outlook, which for us is still billions of years away, the fundamental building blocks of our universe appear in a totally new light.

a. The Dissolution of Matter

Most of the objects we encounter remain the same regardless of how we look at them. For instance, if we examine a rock in a laboratory, the rock remains a rock whether we look at it during the day or at night, with our naked eyes or through a magnifying glass. If we change the actual form

1. Friedrich Engels, *Dialektik der Natur,* in Karl Marx and Friedrich Engels, *Werke* (Berlin: Dietz, 1968), 20:327. A still reliable and informative introduction to the issues touched on in this chapter and the next, especially in their historical context, is provided by Sir William Cecil Dampier, *A History of Science and Its Relations with Philosophy and Religion,* postscript by I. Bernard Cohen, 4th ed. (Cambridge: University Press, 1966).

of an object, for example, if we grind our rock to fine sand, there still remains an object (sand) which is equivalent to the earlier object (rock) plus the energies involved in the transformation process. An object always remains an object. At the most it enters the scene of investigation in a different gown.

Nonetheless, in his second antinomy of pure reason the German philosopher Immanuel Kant showed that our reasoning concerning an absolute object results in a basic conflict.[2] On the one hand, he argued, we can assume that we can divide every composite object into parts, and these parts again into smaller parts, and continue infinitely with this dividing process, obtaining smaller and smaller parts. With equal right we could assume that this process ends at a point when we have reached the ultimately smallest and hence indivisible parts. Kant concluded that both assertions are logically possible and that we cannot decide through experience or perception which one is right. When scientists investigated the realm of the microcosmos, they encountered the same antinomy in a very baffling way. While in our visible world we are able to divide the objects of investigation into smaller and smaller parts, this does not seem to work in the microcosmic world.

In the microcosmic world we first come to molecules and atoms and then to the so-called elementary particles, for instance the quarks, which belong to the fundamental building blocks of matter. Yet one should not consider elementary particles to be analogous to common objects. Though these particles have an electric charge, angular moment, mass, magnetism, and other characteristics of "normal" objects, one cannot attribute to them any extension. In its most minute entities, matter seems to have evaporated to energy bundles. This would also correspond to Albert Einstein's discovery: $e = mc^2$. This equation asserts that mass can be converted into energy, a conversion that takes place, for instance, in atomic reactors, in which fissible uranium decays into other elements while discharging energy. Since Einstein had expressed with this formula a universal relation, it can also be applied to the combustion of coal or wood with the aid of oxygen into carbon dioxide, whereby energy is released in the form of light and heat. In these common combustion processes so little energy is released, however, that

2. For the following, cf. Immanuel Kant, *Critique of Pure Reason* (A.434ff.; B.462ff.), trans. N. K. Smith (London: Macmillan, 1963), 402ff.

in contrast to processes in atomic reactors the decrease of mass cannot be measured. Nevertheless, we do not encounter indestructible objects which one can divide at will, but objects which can at least in part be transformed into energy, a process through which the amount of matter decreases.

The dissolution of the absolute object also led to the uncertainty relation, first formulated by Werner Heisenberg (1901-76).[3] He discovered that in the microcosmic realm the "product" of the uncertainty with which the velocity and location of a particle can be measured never falls below a certain value and is likely to be higher. The problem expressed with this relation can be exemplified roughly in this way: According to Bohr's model of an atom, we can compare an atom to our planetary system. Some planets we can see with the naked eye and others with telescopes. Through such observation we have no problems determining the exact location and velocity of the planets. The reason for this is that planets emit light or reflect light which is then caught by our eyes. The interaction established through these light rays between us as observer and the observed matter is too small for us to influence the velocity or the location of the planet merely through observation, though it is certainly larger than zero.

When we come to the microcosmic realm, however, we cannot observe the electrons which encircle the nucleus of an atom at very high speeds or any other elementary particles with our naked eye. Even the largest microscopes are of no help. We need to use rays, such as light rays or elementary particles whose "size" comes very close to the electrons we want to observe and must "touch" them in the process. Even the tiniest rays we could use are not so small that the object to be observed is not disturbed. This means that through our observation we considerably change both course and velocity of the object to be observed. Heisenberg therefore concluded that though we can determine whether we want to observe more closely the location or the velocity of a particle, the "product" of the uncertainty can never go below a certain limit. In other words, we can determine what facet we want to see more precisely, but we cannot obtain a complete picture of the object of our investigation at one and the same time.

3. For the following, cf. Werner Heisenberg, *Physics and Beyond: Encounters and Conversations,* trans. Arnold J. Pomerans (New York: Harper Torchbooks, 1971), 77ff.

Perhaps we can illustrate by comparing this situation to a snowball fight in the dark. If someone is hit with a snowball, we can approximately determine the location of the person by the surprise cry, but it is difficult to determine from the cry the velocity of the ball that hit the person. If we would throw another snowball into the same direction, we could not be sure that we hit the same person since the former hit does not give us any clues whether the one who was hit will stay in the same location or will move. This is approximately the situation we encounter in the subatomic realm. We can determine certain features of an object under investigation by conducting experiments, but we are unable to know what happens from one experiment to the next. We can, for instance, locate a certain electron at one time here and another time there, but we do not know how it went from here to there.

Even if, hypothetically speaking, we were someday to discover rays which would hardly influence the speed or the location of the object to be observed, such a discovery would not be of much help since these new and very "tiny" rays would then be the fundamental building blocks of our universe. We would have moved, for instance, from electrons to quarks. Then we would have to determine their fundamental characteristics, and to do that we would again have to use rays, and their "magnitude" would not be very different from that of the object to be observed.

The last phenomenon we want to mention in this context is the duality of light. The question about the nature of light has aroused the curiosity of scientists for a long time. Isaac Newton (1642-1727) had already suggested that light consists of certain particles. This idea can best be substantiated through the so-called photoelectric effect. If the light emitted from a certain source were to consist of particles, they would hit a surface with a certain impact. It can be shown that the impact of light "rays" is strong enough to eject electrons from the matter "hit" by light. For instance, if a metallic plate, such as a zinc plate, is attached to a negatively charged electroscope and is exposed to light rays produced by an electric arc, the spread leaves of the electroscope will gradually collapse. This indicates that the electric charge has been lost through emission of (negatively charged) electrons on the metallic plate. When we decrease the amount of light or use a metal that is less chemically reactive, it will take longer for the charge to disappear. If we decrease the frequency but not the amount of light below a certain point, no electrons will be ejected from the metal plate. Finally, if we use a positively charged electroscope, little or no change will be observed, since the electrons emitted from the plate are attracted back by the positive charge of the electroscope. On the basis of these observations, Albert Einstein concluded in 1905 that light consists of photons (particles having no mass), and that the energy of each photon is proportional to the frequency of light. Thus the assumption seemed to be right that light consists of small energy "particles" that can even eject electrons from matter.

However, other observations with the phenomenon of light lead to the conclusion that light consists of waves. For instance, when we send a light beam from a light source through a small slit behind which a screen is mounted, we observe that on the screen the pattern of light is not rectangular as we would have expected if the light only consisted of corpuscles or of particles. The picture on the screen resembles an imaginary frozen surface of a pond after we have thrown a stone into the water; it looks like a picture of concentric rings. When we have two slots parallel to each other and send light through them, we observe a phenomenon analogous to that caused on a pond if two stones are thrown simultaneously into it. The two ring patterns overlap and result in alternating dark and light bands, dark where the two "systems" cancel each other out and light where they reinforce each other. Thus we conclude that light consists of waves.

How can we reconcile these conflicting assumptions? After all, a particle cannot be a wave and vice versa. Scientists tried hard to find a solution but with no lasting success. Finally the Danish physicist Niels Bohr (1885-1962) suggested that we assume that light is *neither* corpuscle nor wave but has features of both, features that complement each other but can never be produced at the same time with the same experiment.[4] The observer's choice of experiment will determine whether it will be possible to observe the effects of light as corpuscle or as wave.

This duality, or complementarity to be more precise, is not only characteristic of light but also of electrons and other elementary parti-

---

4. Cf. Louis de Broglie, *Matter and Light: The New Physics,* trans. W. H. Johnston (New York: Dover, 1946), 111-19, who acknowledges "a far-reaching symmetry between Matter and Light as far as their dual character (waves and corpuscles) is concerned" (157).

cles. They produce both interference phenomena and photoelectric phenomena. Thus we can neither actually visualize the ultimate building blocks of our universe nor univocally determine what they are. "We have to be extremely careful with images in quantum mechanics!"[5] Depending on our method of investigation, these "objects" appear to us in certain ways, but their actual "essence" is beyond these concretions. This means that matter is not irrevocably given but fundamentally relative. Contrary to classical physics, the state of a physical system is not just a catalogue of actualities that is simply so, "but is also a network of potentialities."[6]

b. The Relativity of Space and Time

Immanuel Kant already cautioned in his *Critique of Pure Reason:* those "who maintain the absolute reality of space and time, whether as subsistent or only as inherent, must come into conflict with the principles of experience itself."[7] Kant himself assumed that space and time are forms of outer intuition, they are means with which we perceive an object matter and are not constituent of the object matter. With this definition he rejected the notion that space and time are simply "boxes" within which we locate the objects of our perception.

Later scientific discoveries verified Kant's notion that space and time are not independent entities. By defining space and time as forms of outer intuition, Kant, however, turned them into part of the conceptual apparatus of the observer. This means that Kant took a position opposite to Einstein, who first postulated and then even proved that space and time must be related to the object matter and cannot be perceived apart from it. Already 1,500 years before Einstein, the early church theologian Augustine was not far from this insight when he claimed that to ask what happened before the creation of our world does not make

5. So Alain Aspect, "Wave-Particle Duality: A Case Study," in *Sixty-Two Years of Uncertainty: Historical, Philosophical, and Physical Inquiries into the Foundations of Quantum Mechanics,* ed. Arthur I. Miller (New York: Plenum, 1990), 58, concluding remarks on the outcome of the conference documented in this volume.

6. So Abner Shimony, "Some Comments and Reflections," in *Sixty-Two Years of Uncertainty,* 309, in his instructive paper.

7. Kant, *Critique of Pure Reason* (A.40; B.57), 80f.

sense, because we perceive time only in such a way that we observe a change of events. "There would have been no time if there had been no creation to bring in movement and change."[8] Because nothing existed before the creation of the world, there also was no time. Space and time presuppose matter, or at least a configuration of objects, while empty space would be similar to a hole with nothing around it.

In his theory of relativity Einstein laid down the theoretical foundation for our new understanding of space and time. Einstein postulated in 1905 in his special theory of relativity that the velocity of light is always constant relative to an observer. This means that light emitted from a source does not change its velocity regardless of whether the source or the observer is moving. This seems contrary to all moving objects of which we know. For instance, if a rocket is moving away from an observer with a velocity v and a bullet is fired from a rocket with a velocity u in the direction the rocket is moving, the resulting speed of the bullet for the "stationary" observer is not $v + u$, as might be expected. Einstein made it clear that simply adding the velocities or subtracting them results in an approximate value, a procedure which can only be used with relatively small velocities. When the velocities approach the speed of light (c), we must be more careful in computing the resulting speed by resorting to an equation which prevents any two velocities ensuing in a speed higher than that of light:

$$V = \frac{v+u}{1+ {vu}/{c^2}}$$

This equation also pertains to lower velocities, but there it can be simplified to a level of merely adding the velocities.

The principle that the velocity of light is always constant relative to an observer, together with a second principle of relativity, that all motion is relative, caused a radical reorientation in our concepts of space and time. Before this discovery it was commonly accepted that the whole universe was filled by some kind of world ether in relation to which everything else could be defined in the world, like the chairs in relation to the walls of a room. However, Einstein abandoned the idea of

8. Augustine, *The City of God* (11.6), trans. Henry Bettenson, introduction by David Knowles (Harmondsworth: Penguin Books, 1972), 435.

a world ether and came to the conclusion that there is no fixed world center. We can only define something which is moving by defining its motion in relation to the motion of something else. For instance, when we drive on a road, we assume that the road is at rest. But actually the road is part of the earth, and the earth is once annually moving around the sun and once each day revolving around its axis. The sun is rotating around its axis as well as moving within the Milky Way. The Milky Way itself is moving, and so on. Thus everything is in motion, nothing is at rest, and the idea of a fixed world center is a wishful dream.

The correlative motion of bodies not only determines their respective location but also their extension, mass, and even their time. We might notice while flying in a jet plane that we seem to get heavier when the jet accelerates and takes off. Yet what we feel is not an actual increase in mass, but the effects of acceleration on the inertia of our body. If we would travel at speeds very close to that of light, however, an *actual* increase in mass would take place which would be proportional to the velocity of the body. If a body were to reach the speed of light, its mass would have increased infinitely. This increase of mass can, for instance, be observed in particle accelerators. There particles are charged with a huge amount of energy so that they accelerate close to the speed of light. For instance, the speed of electrons in a particle accelerator can be increased so much that their relative mass becomes ten thousand times that of their mass at rest. If such an electron would be a rocket, so that we could directly observe the accelerated object, we would observe an increase in mass and a corresponding shortening of the object. Yet this decrease in extension again is relative, because it can only be observed by someone who moves with the relatively slower speed than the object to be observed. As soon as we are in the rocket itself, we would not notice a decrease of extension. Yet the space between us and objects outside the rocket would steadily decrease, and if we would obtain the speed of light the difference would be zero; this means that if a body is increased with the speed of light, it would be a mere "energy bundle" without extension.

Since time can only be measured with reference to objects and since objects are relative in speed to their extension and increase in mass, it is a matter of fact that changes also impact the time related to that object. For instance, if we were to leave the earth in some imaginary jet with a speed close to that of light, all our body functions would slow down considerably. We would not notice this, however, since all the other gadgets in our jet, including all timekeeping devices, would slow down accordingly. But a stationary observer on the earth would notice a retardation of our aging, perhaps once we had returned from a long interstellar trip, traveling all the time with extremely high speed. While such retardation of the aging process due to the relativity of time may sound like science fiction, it has actually been observed that timekeeping devices, when highly accelerated, slowed down in that function relative to a stationary observer.

Since time is also relative, the question arises whether simultaneous events are possible.[9] While we cannot deny that events occur at the same time, it is impossible to observe their simultaneity because light always takes time to reach an observer from the objects or events he wants to observe. He always sees things of the past; for him they may have happened at the same time, but they actually occur at different "local" times.

In summary we can state that scientifically speaking space, time, and matter have no absolute characteristics — only relative. There are no unchangeable points of reference which would help us orient ourselves. Space and time are so closely connected with matter that they cannot exist independently from it. But even the "essence" of matter remains hidden from our eyes. Matter can be transformed into energy and it can appear as either corpuscle or wave. It can even increase or decrease depending on its relative speed. Of course, we can still rejoice in the infinite space of the sky or the eternal roaring of the sea when its waves crash against the shore. But in either case we should remind ourselves that the terms "infinite" or "eternal" have only metaphoric value. They describe a finite reality, an entity with boundaries. The elementary building blocks of our universe are only of relative character.

c. The Cause-and-Effect Sequence (Determinism)

In our considerations of the universe and of life, we have always referred to a definite cause-and-effect sequence through which we can recon-

---

9. Cf. the enlightening remarks by Albert Einstein, *Relativity: The Special and the General Theory: A Popular Exposition,* trans. Robert W. Lawson (New York: Crown, 1961), 25ff. and 149, which also presents an excellent introduction to many other problems of Einstein's theory.

struct the early history of our world and of life and which also allows us to glimpse into the future of the universe. The French mathematician Pierre Laplace rightly claimed: "Given for one instant an intelligence which could comprehend all the forces by which nature is animated and the respective situation of the beings who compose it — an intelligence sufficiently vast to submit these data to analysis — it would embrace in the same formula the movements of the greatest bodies of the universe and those of the lightest atom; for it, nothing would be uncertain and the future, as the past, would be present to its eyes."[10]

If we connect all available data, so Laplace, then everything is strictly determined, there are no surprises, past and future are equally accessible for us, and everything will take its predictable course. This wishful dream of the nineteenth century, however, was contradicted by scientific progress. Every discovery, from Einstein's theory of relativity to Otto Hahn's fission of the atom, shows that the course of scientific insight is usually not predictable. There are always surprises which lead to an increase in knowledge and, even more, sometimes to a reformulation of basic convictions. Of course, afterward we can adduce the reasons why certain discoveries or reformulations were made, but these are statements after the event has occurred.

If we want to make everything predictable in our world, then we would have to program a supercomputer with all the data of our world. But such computerization is logically impossible, because we can never program all the causes at a given point in time, since our world is not steady but constantly changing. These changes would require a continuous reconsideration of the already programmed causes. Besides this, a total programming would also necessitate a programming of the programming process, otherwise some potential causes for the future course of the world could possibly be omitted. We quickly realize that such a process can never come to its ultimate conclusion. The only relief could come from a "supermind" that is not connected with our world and its cause-and-effect system. Or put in a more old-fashioned way, only God could know all causes and effects in our world.

---

10. Pierre Simon Marquis de Laplace, *A Philosophical Essay on Probability,* trans. F. W. Truscott and F. L. Emory, introduction by E. T. Bell (New York: Dover, 1951), 4. For a good critique of Laplace's position, cf. Georg Hendrik von Wright, *Causality and Determinism* (New York: Columbia University Press, 1974), 116-19.

Another argument against a stringent causal system comes from the structure of matter itself. In viewing the duality of matter we notice that we can determine whether matter will appear to us in its corpuscle or in its wave character. Matter itself does not determine its characteristics, but we choose between the two possibilities. When we investigate the causal nexus in the subatomic structure of matter, we discover an even more surprising indeterminism.

One of the best-known phenomena of radioactive materials is their spontaneous decay. Radioactive substances disintegrate into substances of less atomic mass whereby part of their former mass is converted into energy. For instance, if we have one kilogram of the radioactive carbon isotope C-14, then according to the radioactive period of 5,730 years for C-14, we can determine that exactly half the material will have decayed after that period. Yet if we have only one radioactive C-14 atom, then we can never precisely predict when it will decay. We know that the higher the intensity of the radioactivity of a specific substance, the shorter the lives of its atoms. Therefore we can determine the mean time of each radioactive substance when half of its substance will be decayed (the so-called radioactive period). But we can never know for sure when a certain radioactive atom will decay. Of course, we could say this is similar to the average life expectancy of a population. We can always predict how long the citizens in a certain country will live on average, but not at which age a certain citizen of that country will die. Of course, we could get the person's health checked and perhaps arrive at a somewhat reliable prediction. But this method will not help with an individual radioactive atom, because an atom does not get sick or decrease in vigor. It is fully "alive" and then, from one moment to the next, it decays without predetermination and without any obvious cause.

For many physicists this spontaneity was extremely uncomfortable. For instance, Einstein mused: "The theory [of quantum mechanics] says a lot, but does not really bring us any closer to the secret of the old one! I, at any rate, am convinced that He is not playing at dice."[11] With this statement he wanted to assert that without sufficient cause there is no effect (meaning the decay of an atom). But this also implied that perhaps

11. Einstein in his letter of December 4, 1926, to Max Born, in *The Born-Einstein Letters,* correspondence between Albert Einstein and Max and Hedwig Born from 1916 to 1955 with commentaries by Max Born, trans. Irene Born (London: Macmillan, 1971), 91.

there were not initial identical states for the two atoms, one of which decayed while the other did not. One surmised that there must be different — though at present still unknown — initial conditions for both atoms. This means one started with hidden variables which, provided their existence could be proved, would again restore a deterministic causal nexus in physics. If the initial value of the hidden variables could be known, one could predict which of the two atoms would decay. This theory of a hidden variable basically resembles the theory of a continuous creation introduced by Fred Hoyle. By introducing unprovable initial conditions Hoyle wanted to escape from a reorientation of our understanding of reality, meaning that the world indeed has a beginning. Yet a more progressive way of thinking seems to gain the upper hand, especially such as that advanced by Bohr in Copenhagen and Heisenberg in Göttingen. The so-called Copenhagen interpretation of quantum mechanics emphasizes that a theory can only be based on what can be observed and experimentally measured. Therefore one must say farewell to the idea of hidden variables which cannot be measured. Indeterminacy is not only a temporary solution which helps us to overcome embarrassing facts, but it is foundational for the structure of reality.

Quantum mechanics was named by its discoverer Max Planck (1858-1947). As early as 1900 Planck assumed to everyone's surprise that energy is not infinitely divisible. He claimed that it always appears in one or a multitude of energy bundles or quanta.[12] When an atom changes its energy level, it either absorbs or emits one or several of those quanta. Since electrons thereby jump from one energy level to another, these changes are called quantum jumps. Again we are confronted with non-causal events; we do not know when a certain electron will jump and which energy level it will assume afterward. Heisenberg was the first to assert that quantum mechanics excludes a deterministic structure of reality. Yet he conceded that "quantum theory always enables us to give full reasons for the occurrence of an event after it has actually taken place," though we cannot predict it with absolute certainty in advance.[13]

Of course, we could cut off the whole discussion by claiming that in

12. Cf. Werner Heisenberg, *Physics and Philosophy: The Revolution in Modern Science*, introduction by F. S. C. Northrop (New York: Harper Torchbook, 1962), 30f.

13. Werner Heisenberg, *Philosophic Problems of Nuclear Science*, trans. F. C. Hayes (London: Faber and Faber, n.d.), 51.

our everyday life we are never confronted with quantum jumps or with the radioactive period of atoms.[14] One could also doubt whether elementary particles have an objective existence such as houses and people do, since we can never directly observe them. Yet we must remember that we can work very well with them to describe the atomic realm which provides also the basic structure for the macrocosmos in which we move. This atomic world is very close to us. We realize this very quickly when we consider that quantum jumps are involved in all polymerization, crystallization, and melting processes. These processes then express themselves in the structural changes of the final product. Though, for instance, the crystalline structure of cast steel conforms to strictly determined patterns, the subatomic starting point of each of its crystallization processes is due to quantum jumps. When a piece of metal breaks or ruptures, the course of breaking or rupturing conforms mostly to the fixed crystalline structure, but in part also to quantum jumps that determine the exact micromolecular point from which the breaking starts. This means that in warfare a splinter of an exploding grenade barely missing a civilian walking down the road is difficult to explain as a strictly predetermined event.[15] The same indeterminacy must be recognized, for instance, in the blowout of a tire which did not seem to be worn out and which led to the deaths of the car's passengers. We could cite many other situations in which the effect of undetermined quantum jumps on our everyday life becomes evident. They play their role not only in fateful occasions, but in such common events as the boiling of water or in the use of household electric current.

14. For the considerable discussion among scientists and philosophers of science concerning the nature and the role of uncertainty, see the excellent summary of William A. Wallace, *Causality and Scientific Explanation,* vol. 2, *Classical and Contemporary Science* (Ann Arbor: University of Michigan Press, 1974), esp. 163-308. The basic issue at stake here is whether "the uncertainties are also objective in the sense that they characterize matter or reality and not merely man's knowledge of such reality" (cf. Wallace, 307). Yet even if we concede that uncertainty is only a characteristic of our knowledge of reality and could some day be replaced by a newly gained deterministic view of reality, what would hinder us from assuming that at a still more distant point in history another indeterministic view of reality might emerge? In other words, the transition from the once-assumed all-embracing deterministic view of reality to an indeterministic one has irrevocably shaken our confidence in stringent determinism.

15. This impressive example is adduced by Pascual Jordan, *Der Naturwissenschaftler vor der religiösen Frage,* 2nd ed. (Oldenburg: Gerhard Stalling, 1964), 154ff.

Another field which should be mentioned in passing and is not un-related to the phenomena emerging in quantum mechanics is chaos re-search. Ideally the main maxim of science is its ability to relate cause and effect. Yet there are natural phenomena that appear to be rather dif-ficult to predict, as we see in weather forecasts. Yet chaos also appears in "the behavior of an airplane in flight, the behavior of cars clustering on an express way, the behavior of oil flowing in underground pipes."[16] As scientists began to look around, they saw chaotic systems everywhere, both in nature and in our own civilization. As we know from weather forecasts, however, chaotic systems are not chaotic, that is, totally un-predictable. There is a degree of irregularity that remains constant over different scales, so that one can talk about a "regular irregularity."[17]

One can demonstrate this phenomenon, for instance, with a Ping-Pong ball which hits a table which also is moving up and down, thus in-creasing the upward and downward movements of the ball. If the table vibrates at a low frequency, the movements of the table and the ball can synchronize. Each time the table moves upward, the ball will then also bounce upward. If the frequency increases, the ball too will jump higher, but only to a certain limit can it synchronize its movements with that of the table. There are two alternatives: (1) The ball can establish a new synchronization. But then only after every second or third upward movement of the table will the ball bounce back, because it needs much longer than the table to conduct its upward and downward movements. (2) The other possibility would be that the ball does not establish a syn-chronization, but hits the table in irregular and seemingly accidental in-tervals. This reaction is also determined, but it is irregular and in this sense chaotic. Through a simple differential equation we can exactly calculate when the next contact of ball and table occurs and how fast the ball will move at the point of contact. Necessary for this calculation, however, are both the exact data of the point of the last contact and the velocity when that contact occurred. In principle the future of the con-tacts between ball and table is completely determined by the past. In practice, however, the exact data of the point of the last contact and the velocity when the contact had occurred will usually only be gathered with some degree of approximation.

16. So James Gleick, *Chaos: Making a New Science* (New York: Viking, 1987), 5.
17. Gleick, 98.

The minute "errors of measurement" which enter into calculations are amplified in the long run "with the effect, that even though the behavior is predictable in the short term, it is unpredictable over the long run."[18] This means that we reach a "predictability horizon" of a system, exemplified here by the Ping-Pong ball and the table. This horizon shows us the limit within which predictability is ideally possible and beyond which we will never be able to predict with certainty. "It has been established, for example, that our predictability horizon in weather forecasting is not more than about two or three weeks. This means that no matter how many more weather stations are included in the observation, no matter how much more accurately weather data are collected and analyzed, we will never be able to predict the weather with any degree of numerical accuracy beyond this horizon of time."[19]

We could argue that a statement which includes the dangerous word "never" is highly controversial. Naturally, if we know the present precisely, we can calculate the future. While this is the case ideally, reality shows a different picture. As the transition point between the still-outstanding future and the already determined past, the present is continuously in flux. Only by moving outside the flux of time can we obtain a complete picture. But we are confined to the present and cannot extricate ourselves from it.

Even if we concede that we can determine the initial conditions with high precision, each physical system is still susceptible to accidental external disturbances which can increase exponentially into a chaotic situation until they void the initial predictions. For instance, it is quite possible that atmospheric movements which can be described by precise equations are in a state of chaos. Therefore there is little hope that the horizon of weather predictions can be expanded ever farther unless we limit ourselves to very general predictions. Yet there are very definite characteristics in a given climate, for instance the annual cycles of temperature and precipitation, which are not influenced by chaos. The desire of Laplace — that a largely predictable universe can be constructed or calculated by considering all initial conditions — cannot be fulfilled.

18. Heinz-Otto Peitgen, Hartmut Jürgens, and Dietmar Saupe, *Chaos and Fractals: New Frontiers of Science* (New York: Springer, 1992), 11.
19. Peitgen, Jürgens, and Saupe, 11.

We should also remember that classical determinism is an abstraction that does not take into consideration that on the microlevel nature is subject to the laws of quantum mechanics and thereby also to the uncertainty principle. This principle shows us that at one specific time we cannot obtain a complete picture of everything. Here chaos theory and quantum mechanics come together, because "one of the lessons coming out of chaos theory is that the validity of the causality principle is narrowed by the uncertainty principle from one end as well as by the intrinsic instability properties of the underlying natural laws from the other end."[20] Yet it would be wrong to conclude that we must abandon causality and determinism. Both are still present but in a weaker or modified form.

Though we discover that much is in flux under the thin veneer of our objective everyday world, that many things are undetermined or only predictable within certain limits, for our own actions we need a pragmatic determinism whose causal connection is not in doubt. For instance, how could an architect design a bridge that can withstand the forces of wind and weather if there were no macrocosmic determinism? Yet some earthquakes have shown that the so-called earthquake-safe structures can be easily destroyed whereas others astoundingly survive. This shows how limited even this pragmatic determinism is. In general, however, determinism serves as a pragmatic basis for our own actions while the substructure of reality corresponds to this determinism only in a very limited way.

The undetermined substructure of reality also becomes evident when we remember how laws of nature are postulated. They are projections from a series of *a* to a totality of *a*. A limited, observable basis serves as the starting point for nonobservable postulations. This does not mean that laws of nature are wrong. Within their limits they are necessary and reliable, and all anticipated events, as soon as they belong to the past, can be checked whether they have fulfilled the predictions of these laws. But these laws are of an irrevocably hypothetical nature. They are not laws according to which events must occur, but laws patterned according to our experience of the way events generally happen. As nature reveals its orderliness in following certain patterns, we should speak rather about *orders* of nature than about *laws* of nature. These laws

20. Peitgen, Jürgens, and Saupe, 14.

have often been mechanistically understood. This is a misinterpretation.[21] Each and every law of nature that exists is not cast in iron but will only be proven in its comprehensive validity at the end of all events. This is also true with regard to the development of the cosmos and the history of life. We simply presuppose that the laws which are valid for us today have been valid everywhere and at all times in the cosmos and are still valid there. For instance, we presuppose that the velocity of light has never changed in the billions of years, and even in the most distant quasars the spectral lines mean the same as they do for us on earth.

While at first glance everything seems to be structured in a causal mechanistic way, we soon realize that this is not the case. The causal picture is the view we get when we look at the surface of our world. But in penetrating its surface structure, we see that its very foundation is undetermined. Whether we consider the infinity of space and time, the eternal duration of matter, or the determined causal nexus, we soon realize the relativity of these absolute-sounding attributes. Even if we must use for our everyday world such "absolute" data as parameters for orientation, we should not forget that they are only of relative significance. They cannot endow our life with ultimate orientation and foundation. Such a foundation must be found beyond the physical, in that which traditionally has been called the metaphysical.

21. Cf. Rudolf Carnap, *Philosophical Foundations of Physics: An Introduction to the Philosophy of Science*, ed. Martin Gardner (New York: Basic Books, 1966), 207, who writes in his accurate description of the prevailing notion of causality in nature: "Perhaps it would be less confusing if the word 'law' were not used at all in physics. It continues to be used, because there is no generally accepted word for the kind of universal statement that a scientist uses as basis of prediction and explanation. In any case, it should be kept clearly in mind that, when a scientist speaks of a law, he is simply referring to a description of an observed regularity. It may be accurate, it may be faulty. If it is not accurate, the scientist, not nature, is to blame."

Part III
Regaining a Christian Faith in Creation

The development of modern industrial civilization and the rapidly occurring and epoch-making discoveries in the natural sciences have contributed to the coronation of the natural sciences as the new queen of all knowledge. Science has attained, for many people, this position while theology has been increasingly relegated to a nonscientific and largely private enterprise. Even in the USA, a country initially settled by people who wanted to live their religious life without the tutelage of kings and princes, the separation between state and church has led to an increasing privatization and marginalization of religion.

The controversy concerning materialism in the 1850s in Germany and the clash of Charles Darwin with the religious authorities (particularly Bishop Wilberforce) are symptomatic of the rather militant approach of the scientific mind-set to traditional religious beliefs in Europe. The progressive spirit of most Americans, however, saw no need to malign science or to perceive it as something that could potentially come in conflict with their religious beliefs. God gave them this new world and therefore anything new — scientific or otherwise — was yet another sign of God's goodness. The Harvard naturalist Asa Gray, in his review of Darwin's treatise *The Origin of Species,* was characteristic of this spirit. Gray pointed out that Darwin's theory of evolution was not a denial of religion, but a scientific theory substantiated on scientific grounds and therefore to be refuted only on these grounds. He also emphasized that Darwin's theory did not diminish God's creative activity. Interpreted theistically, it even enhanced our understanding of the magnitude of divine creation. This would indicate that there is an immediate rapport between the findings of science and the tenets of the Christian faith. Yet when we look more closely, we detect a remarkable silence on the theological side. Neither on the Continent nor in America was there an Asa Gray among the theologians proper who immediately introduced Darwin and his theories to other theologians.

4. The Gradual Rediscovery
of the Created Order

Andrew D. White (1832-1918), in *A History of the Warfare of Science with Theology in Christendom* (1897), mentions "the myriad attacks on the Darwinian theory by Protestants and Catholics."[1] Richard Hofstadter (1916-70) conveys the same sentiment when he says: "The last citadels to be stormed were the churches."[2] Frank Hugh Foster (1851-1935), in his posthumously published book *The Modern Movement in American Theology* (1939), comes much closer to the truth when he suggests: "In strict accordance with its own principles, the appearance of evolution on the theological stage and the perception of its importance for the philosophy of religion was a very gradual affair."[3] Indeed, initially theologians paid very little attention to science and were either content to show that science would not hurt theology or that theology was something quite different from science. Only relatively late did a negative backlash against science arise, and especially against evolution. This was mainly fostered through contact with Continental theologians. After that episode, in order to hold on to one's own religious beliefs, one withdrew more and more from the scientific world.

1. Andrew D. White, *A History of the Warfare of Science with Theology in Christendom*, 2 vols. (New York: D. Appleton, 1897), 1:78.
2. Richard Hofstadter, *Social Darwinism in American Thought*, rev. ed. (New York: George Braziller, 1969), 24.
3. Frank Hugh Foster, *The Modern Movement in American Theology: Sketches in the History of American Protestant Thought from the Civil War to the World War* (New York: Fleming H. Revell, 1939), 38.

a. The American Way: Indifference — Caution — Embrace — Neglect

Initial Reaction in Periodical Literature

When we take a quick look at theological periodical literature, we find one of the first times Darwinism is mentioned is in 1863 in a brief review of Dana's *Manual of Geology* in *Bibliotheca Sacra*. There the reviewer stated: "The supporters of Darwinism will find but little comfort in this volume. Professor Dana fully believes in the creation of successive races of animals and plants at different periods."[4] In 1867 a long article in two installments appeared in *Bibliotheca Sacra:* "The Relations of Geology to Theology" by C. H. Hitchcock of New York City. With reference to the works of James Dwight Dana (1818-95) and because of the scanty scientific evidence for proving the evolutionary theory, Hitchcock rejected the idea that one species developed from another, especially if applied to humanity. But then he stated:

> Granting the truth of Darwinism, or any judicious modification of its principles, the foundation of our argument is rather strengthened than destroyed. The theory of development may be used like the nebular hypothesis. The latter was devised by La Place to sustain atheism, but after being avoided by theologians as long as possible, has been generally adopted by them, and is turned against its original friends. Hence we say to the development school, go on with your investigations, and if you succeed in establishing your principles we will use your theory for illustrating the argument for the existence of God.[5]

This is certainly not an endorsement of evolutionary thought. But such was not given by all the leading scientists either, neither by Dana nor by Louis Agassiz (1807-73), who remained cold toward Darwinian thought since that seemed to imply for him continued progress, a thought he was unwilling to affirm. Yet Hitchcock did not slam the door to the acceptance of evolution. He rather encouraged science to continue its research, in the assurance that once the new theory had sufficient credibility it would be amenable to the Christian faith.

4. *Bibliotheca Sacra* 20 (January 1863): 222.
5. *Bibliotheca Sacra* 24 (April 1867): 371.

When we turn to the *Baptist Quarterly*, we find the first treatment of evolutionary thought in 1868, in its second volume, with an article, "Development versus Creation," by Heman Lincoln of Providence, Rhode Island. In his extensive review of Spencer and Darwin, Lincoln came to the conclusion that "if the theory of development cannot be accepted as an established scientific *law*, it is at least entitled to favor as a scientific *hypothesis*, which explains many curious riddles in the organic and animal kingdoms. We concede readily that it relieves some perplexities, and explains satisfactorily some phenomena, and may claim attention as an ingenious hypothesis, which, in a future day may possibly unfold into a well-defined law."[6] Thus again the new evolutionary theory was not completely rejected. Lincoln even assumed that if the theory of development should ever be found to rest on a substantial basis of facts, it need not shake one's faith in a divine author of the universe. It may be held also without impairing faith in a true creation, or in the divine government of the world.[7]

In 1874 the *Baptist Quarterly* published an article, "Causes and Final Causes," by F. B. Palmer of Brookport, New York, in which he took Huxley and, to a lesser degree, Darwin to task for removing any design from their theories and especially for ridiculing theology. A review of Charles Hodge's *What Is Darwinism?* in the same year agreed with Hodge's verdict that the exclusion of design from nature is tantamount to atheism.[8] But in the same year the *Baptist Quarterly* also carried an article by Lewis E. Hicks of Granville, Ohio, "Scientists and Theologians: How They Disagree and Why." Hicks showed that scientists were divided into unbelievers, doubters, and believers, but there was no necessary conflict between science and Christianity, only between its doubting or unbelieving adherents.

Looking at the *Methodist Quarterly Review*, we find the first review of Darwin's *Origin of Species* in an extensive 1861 article by W. C. Wilson of Dickinson College, "Darwin on the Origin of Species." This informed review did not deal with theological issues. Siding with Dana, Agassiz, and even Gray, Wilson came to the conclusion that, apart from whatever scientific value the ingenious work of Darwin had, it failed "to re-

---

6. *Baptist Quarterly* 2 (1868): 270.
7. *Baptist Quarterly* 2 (1868): 264.
8. *Baptist Quarterly* 8 (1874): 375.

establish on a scientific basis the often rejected theory of the transformation of species" and soon would be consigned "to its appropriate place in the museum of curious and fanciful specialties."[9]

Henry M. Harman of Baltimore, Maryland, wrote on natural theology in this journal in 1863. He claimed that "the only form of infidelity from which Christianity has anything to fear is the *Theory of Development.*"[10] Yet he rejected Darwin's theories not on theological but on scientific grounds, citing many of the scientists who were opposed to it. He also criticized Darwin for doing away with design in nature, although he admitted that Darwin's theory explained some facts, as any hypothesis would do. In 1865 we find in the same journal an extensive review by John Johnston of Wesleyan University, Middletown, Connecticut, of James Dana's *Manual of Geology.* Johnson hailed it as "an excellent treatise . . . [which] will mark an era in the History of American Geological Science."[11] Clearly in the mid-1860s Darwin's theory was far from being accepted by either scientists or theologians.

The popular *Religious Magazine and Monthly Review* did not publish an article on evolutionary thought until 1871: "Darwin's Descent of Man," signed "G.E.E." The writer tells us that the fright and indignation that existed among so many at the prospect of this book had subsided to a large degree. The reason was twofold: (1) Scientists assured the believing public that Darwin's theory is at no point hostile to or inconsistent with "an unimpaired religious faith in God and Christ and immortality." (2) Many people realized that Darwin's theory "as applied to man, falls so far short of being demonstrated or proved."[12] Nevertheless, the writer called Darwin's book fascinating and recommended it to readers.

When we come to the new series of the *Quarterly Review of the Evangelical Lutheran Church,* published in 1871, we notice in its first volume two review articles dealing with evolution. The first, an extensive unsigned article, "The Theistic Argument from Final Causes," was largely negative. The ideas of Spencer, Huxley, and Darwin were rejected since their work ran contrary to "the accepted principles underlying Natural

---

9. *Methodist Quarterly Review* 43 (1861): 627. Foster's observation that the *Quarterly* "from 1860 to 1880, has no single attempt at a discussion of any theological or scientific bearings of Darwin's work!" is certainly mistaken (cf. Foster, 42).

10. *Methodist Quarterly Review* 45 (April 1863): 183.

11. *Methodist Quarterly Review* 47 (July 1865): 378.

12. *Religious Magazine and Monthly Review* 45 (1871): 502.

Theology and Christian truth." "But," the reviewer stated, "were the entire Development Theory, from the nebular hypothesis of LaPlace to the evolution scheme of Darwin, verified as true cosmogony and science, it would not even then necessarily destroy the evidences of design. It would require the same infinite intelligence to create a universe out of nebular matter and primordial conditions, by the long process of development, as by the direct exercise of creative power. A development theory might be held, in harmony with a certain kind of theism."[13] The reviewer continued to show that the theory had not yet been accepted as science. This seemed to indicate that, though rejected on grounds of novelty and weak scientific backing, the evolutionary theory might in the future be accepted without endangering the Christian faith.

The second article is a review of St. George Mivart's (1827-1900) *The Genesis of Species* (1871). The book, the reviewer said, opted for a thoroughly theistic interpretation of evolution and natural selection. The work was "refreshing" and "entitled to the highest consideration." While the reviewer did not want to follow the author in every detail, he regarded Mivart's general theory as "perfectly consistent with a genuine theism."[14]

The following year we find in the quarterly a long review of *The Descent of Man* by Cyrus Thomas De Soto of the U.S. Geological Survey. De Soto's review reaffirmed that the Darwinian theory was untenable and that natural selection could not be the origin of the species. Yet he conceded "that nothing even in Mr. Darwin's theory, as then put forth, and *a fortiori* in evolution generally, was necessarily antagonistic to Christianity." He even called Thomas H. Huxley (1825-95) "one of the great scientific teachers of the day," but one who wages war against Christianity. In the latter assessment he was certainly not incorrect. While De Soto did not admit that natural selection was *the only* cause of the development of the species, he indeed thought there was a development. But he objected to the notion that all animals could have developed from one primordial form; nor, he argued, could humanity have descended from animals.[15] Thus De Soto opted for a modified evolution without accepting

---

13. *Quarterly Review of the Evangelical Lutheran Church* 1 (April 1871): 182 and 184f.

14. *Quarterly Review of the Evangelical Lutheran Church* 1 (July 1871): 477, 480.

15. *Quarterly Review of the Evangelical Lutheran Church* 2 (April 1872): 213-41, see 221ff.; quotes earlier in paragraph on 241.

the Darwinian theory and arrived at this position from his understanding of Scripture and his knowledge of natural history.

We could continue our review of theological periodical literature of the 1850s and 1870s in much more detail, and it would reinforce our present observations. During that period there were not many outright rejections of evolutionary thought in general. Yet many implied that the Darwinian theory of natural selection had a shaky foundation. The main arguments against the Darwinian theory did not come from theology. Theologians did not conduct a battle between the biblical truth and the knowledge of science. But they drew their arguments from their own scientific knowledge. Quite often they conceded that if the Darwinian theory should be proven correct, it would not pose any threat to the Christian faith since it could be interpreted theistically.

The Fears of Charles Hodge

The most significant and influential attack on evolutionary thought came from Charles Hodge (1797-1878), professor of theology for more than fifty years at Princeton Theological Seminary. He was the leading theologian in his own Presbyterian denomination and, having just published his three-volume *Systematic Theology* (1871), one of the most prominent in the United States. In 1874 he published *What Is Darwinism?* in which he sought to demolish the Darwinian heresy. According to Hodge, Darwin's "grand conclusion is 'man (body, soul and spirit) is descended from a hairy quadruped, furnished with a tail and pointed ears, probably arboreal in its habits, and an inhabitant of the Old World.'"[16] Yet, as Hodge implied, Darwin did not say anything about the human soul. Darwin also would have agreed with Hodge's suggestion: "In using the expression Natural Selection, Mr. Darwin intends to exclude design, or final causes" (41).

Though employing a certain degree of overkill in his argument, Hodge did not want to be unfair to Darwin. He conceded that Darwin explicitly and repeatedly admitted the existence of a creator. But then

---

16. Charles Hodge, *What Is Darwinism?* (New York: Scribner, Armstrong & Co., 1874), 39f. Parenthetical references in the following text are to page numbers from this work.

he criticized him for not saying anything about "the nature of the Creator and of his relation to the world" (27). Hodge seemed to forget that in order to remain credible in his empirical work a scientist must establish the theories without reference to God. This inability to distinguish between empirical and metaphysical arguments became clear when Hodge exclaimed in reference to complicated organs of plants and animals: "Why doesn't he say, they are the product of the divine intelligence? If God made them, it makes no difference, so far as the question of design is concerned, how He made them: whether at once or by a process of evolution. But instead of referring to the purpose of God, he laboriously endeavors to prove that they may be accounted for without any design or purpose whatever" (58).

Like Agassiz, Hodge admitted that God could have made the living beings at once or gradually through the process of evolution. But unlike Agassiz, he did not fault Darwin for advocating evolution. What he rejected was the presentation: evolution was explained in natural terms instead of supernatural ones. By explaining the evolutionary process in natural terms and by natural causes, Hodge implied that Darwin had effectively banished God from the world. It is important to note that Hodge distinguished here between "Darwinism," meaning the denial of any final causes and therefore the explanation of the development of the world without reference to God, and "evolution," referring to the progressive development of the world through God's design (104). He realized that one could affirm evolution without admitting Darwinism.

The reason for Hodge's uneasiness with Darwinism is evident. "God, says Darwin, created the unintelligent living cell . . . after that first step, all else follows by natural law, without purpose and without design."[17] To remove design from nature is tantamount to the dethronement of God the creator. Hodge's verdict: "The conclusion of the whole matter is, that the denial of design in nature is virtually the denial of God. Mr. Darwin's theory does deny all design in nature; therefore, his theory is virtually atheistical; his theory, not he himself. He believes in a Creator."[18] Hodge's evaluation of Darwin culminated in the paradox: "A man, it seems, may believe in God, and yet teach atheism."[19]

17. Charles Hodge, *Systematic Theology*, 3 vols. (New York: Scribner, 1871), 2:15.
18. Hodge, *What Is Darwinism?* 173.
19. Hodge, *Systematic Theology*, 2:19.

Before we reach an assessment of Hodge's position, we must clarify one point: Hodge did not reject evolution in general but Darwin's theory of evolution. What made him react so vehemently against Darwin's theory? We perhaps get a clue when we consider whom he read to confirm his fears concerning the implications of Darwin's theory. We read the names of Alfred Russel Wallace (1823-1913), Thomas H. Huxley, Ludwig Büchner (1824-99), Carl Vogt, Ernst Haeckel, and David Friedrich Strauss (1804-74). For instance, he quoted Haeckel as saying "that Darwin's theory of evolution leads inevitably to Atheism and Materialism."[20] Hodge was familiar with the Continental discussion about Darwin and the antireligious propaganda by people such as Vogt, Büchner, Haeckel, and Strauss. He was afraid that the same might happen in the United States. But his fears were unfounded for two reasons: (1) The evolutionary ideas that came from England were not so much those of Darwin as those of Spencer.[21] (2) Neither Darwin's nor Spencer's theories were simply received in the United States without adaptation. As Hodge perceptively noted, Darwin's most fervent advocate in America, Asa Gray, though an avowed evolutionist, was not a Darwinian. He interpreted Darwin's theory theistically.[22] The same is true of Spencer's philosophy. It was introduced in America through the writings of the philosopher and champion of evolutionary thought John Fiske. In the United States materialists and atheists had no chance of turning evolutionary theory into an instrument that would advance their cause.

There was still another reason for the theistic reception of evolutionary thought in the United States. Most institutions of higher learning which would provide the platform for an intellectual exchange concerning evolution were church-operated or at least in some way affiliated with the church. In England and especially on the Continent, however, they were mostly state-owned and thus provided a more liberal intellectual environment unrestrained by ecclesiastical guidance.

In May 1874 Gray published an extensive review of *What Is Darwinism?* declaring that one should not blame a naturalist for leaving the

20. Hodge, *What Is Darwinism?* 95.
21. Darwin never visited the United States as Spencer had done. On his 1882 visit, Spencer was celebrated and treated like royalty.
22. Hodge, *What Is Darwinism?* 174f.

problems of purpose and design to the philosopher and theologian.[23] Purpose on the whole, Gray asserted, was not denied but implied by Darwin. Gray rightly concluded that Hodge's treatise "will not contribute much to the reconcilement of science and religion."[24] As a result of Hodge's pamphlet many people who had never read a line of Darwin became convinced that Darwin was the great enemy of the Christian faith. But by now the great opponent of evolution, Agassiz, had died (1873); Dana, the leading figure among American geologists, had in the 1874 edition of his *Manual of Geology* endorsed the concept of natural selection; and George F. Wright (1838-1921), then at Andover Theological Seminary, had helped Gray publish his *Darwiniana* (1878).

From Hesitancy to Enthusiasm

That even conservatives had become amenable to evolution can be seen in J. William Dawson (1820-99), who had once supported Hodge and in 1890 stated in his book *Modern Ideas of Evolution as Related to Revelation and Science*, that the current Darwinian and neo-Lamarckian forms of evolution "fall certainly short of what even the agnostic may desiderate as religion."[25] But then he asserted: "Creation was not an instantaneous process, but extended through periods of vast duration. In every stage we may rest assured that God, like a wise builder, used every previous course as support for the next; that He built each succeeding story of the wonderful edifice on that previously prepared for it; and that His plan developed itself as His work proceeded."[26] Evolution was no longer something objectionable as long as it was not Darwinian, that is, proceeding with blind force and blind chance, or Lamarckian, propelled by the impact of the environment.

Even before Dawson, James McCosh, a philosopher-theologian and president of Princeton College (1868-88), had accepted evolutionary

23. *Nation* (May 28, 1874), reprinted in Asa Gray, *Darwiniana: Essays and Reviews Pertaining to Darwinism*, ed. A. Hunter Dupree (Cambridge: Harvard University Press, 1963), 266-82.

24. Gray, *Darwiniana*, 279.

25. J. William Dawson, *Modern Ideas of Evolution as Related to Revelation and Science*, ed. William R. Shea and John F. Cornell (New York: Prodist, [1890], 1977), 226.

26. Dawson, 230.

thought in Hodge's own backyard. McCosh was critical of Darwin's theory, especially of his attempt to attribute the whole evolutionary process to natural selection. He also doubted that humanity should be as closely associated with the animal kingdom as Darwin had claimed. But he confessed: "There are clear indications, in the geological ages, of the progression from the inanimate up to the animate and from the lower animate to the higher. The mind, ever impelled to seek for causes, asks how all this is produced. The answer, if an answer can be had, is to be given by science, and not by religion; which simply insists that we trace all things up to God, whether acting by immediate or by mediate agency."[27]

Here a leading figure of American Presbyterianism declared his acceptance of the Darwinian theory. Yet he was not simply going with the times. As McCosh acknowledged, it had become known "that Darwin was a most careful observer, that he published many important facts, that there was great truth in the theory, and that there was nothing atheistic in it if properly understood."[28] But McCosh was also compelled by an evident pastoral concern:

> I have all along had a sensitive apprehension that the undiscriminating denunciation of evolution from so many pulpits, periodicals, and seminaries might drive some of our thoughtful young men to infidelity, as they clearly saw development everywhere in nature, and were at the same time told by their advisers that they could not believe in evolution and yet be Christians. I am gratified beyond measure to find that I am thanked by my pupils, some of whom have reached the highest position as naturalists, because in showing them evolution in the works of God, I showed them that this was not inconsistent with religion, and thus enabled them to follow science and yet retain their faith in the Bible.[29]

When the review article of George F. Wright (Andover, Mass.) ("Recent Works Bearing on the Relation of Science to Religion. No. V: Some Analogies between Calvinism and Darwinism") argued that Darwinism was the Calvinistic interpretation of nature since it was non-

27. James McCosh, *Christianity and Positivism: A Series of Lectures to the Times on Natural Theology and Apologetics* (New York: Robert Carter, 1871), 63.

28. James McCosh, *The Religious Aspect of Evolution* (New York: Scribner, 1890), vii.

29. McCosh, *Religious Aspect of Evolution*, ix-x.

sentimental, realistic, and to some extent fatalistic, the article was a sign that evolutionary thought had become respectable.[30] This became even more obvious when the most prominent preacher of that time, Henry Ward Beecher (1813-87), finally came out in favor of evolution. In *Evolution and Religion* Beecher declared that "the theory of Evolution is the *working* theory of every department of physical science all over the world."[31] He claimed that it was taught in all schools of higher education and the children were receiving it, since it was fundamental to astronomy, botany, and chemistry, to name just a few. But Beecher insisted that evolution was "substantially held by men of profound Christian faith" and although theology would have to reconstruct its system, evolution would "take nothing away from the grounds of true religion" (19).

The reason for Beecher's confidence regarding evolution was his belief in two kinds of revelation: "God's thought in the evolution of matter" (nature) and "God's thought in the evolution of mind" (reason and religion) (15). Our task is to unite and to harmonize them; and then we will notice that the interpretation of evolution "will obliterate the distinction between natural and revealed religion, both of which are the testimony of God" (20). Beecher was convinced that there could be no disharmony between the God who was active in nature and the God disclosing himself in Scripture. But he even went one step further, a step that eventually evoked protest from the conservative side: he asserted that God disclosed himself as much in nature as in religion. Thus natural religion was revealed religion.

Under Beecher's influence Lyman Abbott (1835-1922), Beecher's successor at Plymouth Church (Congregational) in Brooklyn, New York, joined the ranks of theistic evolutionists and contributed much through his sermons and his journalistic efforts to the idea that Darwinism was acceptable to Protestant thought.[32] In his *Reminiscences* (1915) Abbott confessed that in 1866 he studied Spencer but not Darwin or Huxley since he

---

30. *Bibliotheca Sacra* 37 (January 1880): 76.

31. Henry Ward Beecher, *Evolution and Religion* (New York: Fords, Howard & Hulbert, 1885), as reprinted in part in *Evolution and Religion: The Conflict between Science and Theology in Modern America*, ed. Gail Kennedy (Boston: D. C. Heath, 1967), 18. The parenthetical references in the following text are to page numbers from the 1967 edition.

32. So Ira V. Brown, in his interesting study, *Lyman Abbott, A Christian Evolutionist: A Study in Religious Liberalism* (Cambridge: Harvard University Press, 1953), 141.

was not much interested in science.[33] In *The Theology of an Evolutionist* (1897), however, he called himself "a radical evolutionist" or "a theistic evolutionist."[34] We are immediately assured that he reverently and heartily accepts "the axiom of theology that a personal God is the foundation of all life" but that he also believes "that God has but one way of doing things; that His way may be described in one word as the way of growth, or development, or evolution, terms which are substantially synonymous."

While Abbott noticed that all biologists were evolutionists, he also observed that not all were Darwinians; that is, not all regarded the struggle for existence and survival of the fittest as adequate statements of the process of evolution.[35] He understood evolution as the history of a process, and not an explanation by giving causes. Therefore he accepted the aphorism of John Fiske (1842-1901), a popular lecturer and historical writer: "Evolution is God's way of doing things."[36]

By the 1890s evolution had become a universal system and was also applied to the Bible. Here of course, the big problem was how to reconcile the story of the fall with the descent, or rather ascent, of humanity. Abbott discovered that, apart from Genesis 3, the story of the fall played no role in the Old Testament. Even in the New Testament there is no mention of it, except by Paul when he talks of the struggle between flesh and spirit. Abbott found that Paul's description of this struggle was effectively interpreted by "the evolutionary doctrine that man is gradually emerging from an animal nature into a spiritual manhood."[37] Abbott understood Paul to say that sin "enters every human life, and the individual 'falls' when the animal nature predominates over the spiritual."[38] Incarnation is then interpreted as the perfect dwelling of God in a perfect human being. For Abbott Christ lived and suffered "not to relieve men from future torment, but to purify and perfect them in God's likeness by uniting them to God."[39] Christ did not appease God's wrath, he simply laid down his life in love that others might receive life.

---

33. Lyman Abbott, *Reminiscences* (Boston: Houghton Mifflin, 1923), 285.

34. For this and the following quotes, see Lyman Abbott, *The Theology of an Evolutionist* (Boston: Houghton Mifflin, 1898), 9.

35. Abbott, *Theology of an Evolutionist,* 6f. and 19.

36. Abbott, *Reminiscences,* 460, and many other places.

37. Abbott, *Reminiscences,* 459.

38. Abbott, *Theology of an Evolutionist,* 186.

39. Abbott, *Theology of an Evolutionist,* 190.

Like Beecher, Abbott was convinced that God, dwelling in the world, spoke through all its phenomena. Suddenly evolution had not only become acceptable to Christian faith but had also become the tool with which to interpret the Christian faith and religion in general.[40]

The Reception of Darwinism

With relative ease Darwinism became accepted in America in a thoroughly theistic fashion. This was different from the bitter struggle over Darwin between the freethinkers and the conservatives in Germany, a struggle that continued well into the twentieth century. But actually it was not Darwin and his theory of natural selection that became accepted but Spencer and his cosmic theory of an all-encompassing evolutionary process and of the survival of the fittest. For a young and expanding country like the United States, it was only fitting that the biological theory of Darwin became an appendix to the social, economic, and philosophical theory of Spencer.

The social Darwinism, or rather Spencerianism, of William Graham Sumner (1840-1910), John D. Rockefeller (1839-1937), and Andrew Carnegie (1835-1919) is still with us when those on welfare are classified as lazy, or when, regardless of our calls for hidden and overt government support, we find that free enterprise is the best economic system, or when competition is believed to supply us indefinitely with oil and natural gas. According to its own principles, this kind of Darwinism will have to modify itself either through pressure from outside or from within; or, if it does not change, it will be modified through the collapse of the socioeconomic system. But this Darwinism, widely advocated by the so-called political conservatives, did not make much of a stir in theology. It has therefore been widely neglected by theologians since theology, being usually exercised by members of the socioeconomic establishment or the "fittest," benefits from it.

There is also a liberal Darwinism, which is perhaps even causally related to the first kind. This optimistic evolutionism considers devel-

---

40. Cf. Washington Gladden, *Who Wrote the Bible? A Book for the People* (Boston: Houghton Mifflin, 1891), in which he attempted to demonstrate that the Bible had a "natural history" as well as a supernatural one.

opment and evolution as God's way of doing things. As William James (1842-1910) perceptively noted, "the idea of a universal evolution lends itself to a doctrine of general meliorism and progress which fits the religious needs of the healthy-minded so well."[41]

It is interesting that James, who once learned and taught with Fiske at Harvard, pointed out the shortcomings of this new optimistic religion of nature (the form in which Darwinism was introduced by Fiske, Beecher, and Abbott). James criticized it for its attempt to explain evil away instead of seeing it as an intrinsic part of existence. He correctly stated: "The method of averting one's attention from evil, and living simply in the light of good is splendid as long as it will work."[42] And it did work as long as America was expanding and was still unaware of its boundaries and limitations. But with World War I and the Great Depression, things appeared in a different light. Then many people discovered, as James did in 1902, that Christianity was not synonymous with the gospel of the essential goodness of humanity and of eternal Darwinian (better: Spencerian) progress. They remembered that Christianity was essentially a religion of deliverance, that we are called to die before we can be born again into real life.[43] People felt betrayed by the unjustified evolutionary optimism, and some demanded that evolutionary theories be outlawed altogether.

The course of events might have been considerably different if evolutionary thought had not made its strongest impact on the American mind through Spencer and his interpreter Fiske, who saw in evolution how God acted in nature and history. If it had been through Darwin and his interpreter Gray, who confessed himself to be "a Darwinian, philosophically a convinced theist, and religiously an acceptor of the 'creed commonly called the Nicene,' as the exponent of the Christian faith," both social Darwinism and the conservative backlash might have been avoided.[44]

---

41. William James, *The Varieties of Religious Experience: A Study in Human Nature* (New York: Random House, Modern Library, n.d.), 90. Cf. also Edward A. White's penetrating study, *Science and Religion in American Thought: The Impact of Naturalism* (Stanford: Stanford University Press, 1952), 4-8, where White emphasizes the influence of William James and Reinhold Niebuhr in the rediscovery of the true significance of the Christian faith against optimistic evolutionism.

42. James, 160.

43. Cf. James, 162.

44. Gray, *Darwiniana*, vi.

We must remember how Darwin was received in America if we want to assess properly the lasting impact of his ideas. Darwin's evolutionary theory was introduced in America in a decidedly theistic framework. Initially this mitigated the possible clash with the tenets of the Christian faith concerning creation and providence. The vast majority of American Protestant theologians initially saw nothing in Darwin's theory that was irreconcilable with the Christian faith, provided the theory was scientifically acceptable and was clad in a theistic framework that maintained a personal God who created and sustained the world. In the wake of the expansion of the new American continent, Darwin's theory was seen as part of Spencer's comprehensive evolutionary theory, which also included socioeconomic aspects. After its initial overwhelming success, this idealistic and speculative system clashed with the reality of radical evil and injustice exhibited in history and society. Failing to distinguish between Spencer and Darwin, more conservative theological minds began to react against evolutionary theory in general; and some wanted to ban it from the earth altogether.

The Social Gospel movement at the turn of the century still accepted evolutionary categories in its attempt to address the social injustices that accompanied the phenomenal expansion of America by emphasizing the social dimension of sin. This is evident in remarks by Walter Rauschenbusch (1861-1918), the most prominent representative of this movement: "Jesus was not a pessimist. Since God was love, this world was to him fundamentally good. He realized not only evil but the Kingdom of Evil; but he launched the Kingdom of God against it, and staked his life on its triumph. His faith in God and in the Kingdom of God constituted him a religious optimist."[45] For him, Jesus took "his illustrations from organic life to express the idea of the gradual growth of the Kingdom. He was shaking off catastrophic ideas and substituting developmental ideas" (220). The evolutionary, forward-reaching and upward-moving process was central to the ideas of social betterment espoused by the Social Gospel. Yet Rauschenbusch also recognized that World War I "has deeply affected the religious assurance of our own time, and will lessen it still more when the excitement is over and the af-

---

45. Walter Rauschenbusch, *A Theology for the Social Gospel* (1917; reprint, New York: Macmillan, 1922), 156. Parenthetical references in this paragraph are to page numbers in this work.

termath of innocent suffering becomes clear" (181). Although the progressive drive was deeply entrenched in the American spirit, there were ominous signs that affairs might not continue as usual. World War I was a relatively short episode for America, since America entered it only at the tail end. But the many thousands of European immigrants pouring into America as a result of the war showed that the victory gained had not solved many problems.

The Conservative Backlash

In America, conservative movements picked up significant momentum in the first decades of the twentieth century. For instance, the temperance movement, interrupted by the internal strife of the Civil War, gained amazing popularity and finally led to Prohibition starting in January 1920. This was celebrated by evangelicals as a major victory against social evils such as poverty and the corruption of morals. A few years earlier the publication of a series of small volumes of essays entitled *The Fundamentals* (1910-15) signaled another breakthrough for the conservative cause. Against the ever growing influence of liberal Continental theologians such as Albrecht Ritschl (1822-89), Martin Rade (1857-1940), and Adolf von Harnack (1851-1930), an influential group of British, American, and Canadian writers presented the conservative stand. In this somewhat uneven series, conservative but scholarly contributions were mingled with dispensationalist articles. The series contained three papers on evolution, including one with the characteristic title "The Decadence of Darwinism."[46] Eventually three million copies of *The Fundamentals,* financed by two wealthy laypeople, were distributed to pastors, evangelists, missionaries, theology students, and active laypeople throughout the English-speaking world. The five fundamentals testified to in these volumes were the inerrancy of the Bible, the virgin birth, the atonement, the resurrection, and the second coming of Christ. While *The Fundamentals* could not stop the liberal trend by rallying the conservative forces, it widened the gulf between the two.

46. Henry H. Beach, "The Decadence of Darwinism," in *The Fundamentals: A Testimony to the Truth,* ed. R. A. Torrey, A. C. Dixon, et al., 4 vols. (1917; reprint, Grand Rapids: Baker, 1996), 4:59-71.

The fundamentalists were determined to stamp out, wherever possible, teachings which appeared to contradict Scripture. Sooner or later this would lead to a clash with the theory of evolution. This clash was even more likely since not everyone was preoccupied with progress. Large numbers of people outside metropolitan centers and places of learning were virtually unaffected in their beliefs and habits by the intellectual and cultural climate of the day. They lived in essentially the same way, in the same world, and with the same beliefs as their pioneer ancestors had. Their conservative mood needed only to be rallied around a common cause, and they could form a respectable force in society.

One such rallying point proved to be the teaching of evolution in public schools. Between 1920 and 1930 some thirty-seven antievolution bills were introduced in twenty state legislatures and passed in several states such as Tennessee, Mississippi, and Arkansas. For instance, fundamentalist groups had become powerful enough in Tennessee to pressure the state legislature in 1925 to adopt legislation making it unlawful to "teach any theory that denies the story of the divine creation of man as taught in the Bible."[47]

The antievolution issue came to a climax when, in the summer of the same year, the high school teacher John Scopes of Dayton, Tennessee, was accused and put on trial for violating the recently passed statute prohibiting the teaching of evolution in tax-supported schools. The trial gained lasting fame since two prominent people took sides in it. On the side of the law was William Jennings Bryan (1860-1925), three-time presidential hopeful and ardent champion of the fundamentalist cause; and on the side of the accused, Clarence Darrow (1857-1938), famous criminal lawyer and militant agnostic who sharply ridiculed biblical literalism. The trial aroused not merely national but international attention and was accompanied by an immense amount of publicity. Although Scopes's conviction in the lower court was overturned by the Supreme Court of Tennessee on grounds that the fine had been improperly imposed, the effect on the general public of the publicity was to discredit fundamentalism. As time passed, fewer and fewer thoughtful people took seriously the categorical rejection of evolution by fundamentalists; and this extreme form of the issue virtually passed from

47. According to Clifton E. Olmstead, *History of Religion in the United States* (Englewood Cliffs, N.J.: Prentice-Hall, 1960), 549.

the American scene. Of course, there are still people today who advocate the teaching of the first chapters of Genesis as an alternative to the teaching of evolution in public schools; but the very fact that they advocate it as an alternative indicates that they assume the biblical creation stories and the theory of evolution actually cover the same ground. They have not really discerned the difference between the scientific or physical level of reality and the spiritual or metaphysical level.

Benign Neglect

Langdon Gilkey (b. 1919) pointed to the obvious dichotomy of modernity when he metaphorically described humanity as a "helpless patient in the backless hospital shift and yet as mighty doctor in the sacral white coat."[48] The large spectrum of twentieth-century conservative or neoorthodox theology has never even intended to relate carefully the scientific claims concerning evolution to the spiritual claims of creation. When we briefly look at the most prominent representatives of neoorthodoxy in America, Reinhold Niebuhr (1892-1971) and H. Richard Niebuhr (1894-1962), we do not find any reference to evolution in their major writings. For instance, Reinhold, in his seminal work *The Nature and Destiny of Man* (1941), makes no mention of evolution, Darwin, or Spencer. Referring to the modern view of humanity, he briefly describes the idea of progress as one which, after eliminating the Christian doctrine of sinfulness, relates "historical process as closely as possible to biological process and which fails to do justice either to the unique freedom of man or to the demonic misuse which we may make of that freedom."[49] Similarly in his essay "The Truth in Myths" (1937), he refers to the myth of creation and claims that one ought to distinguish between what is "primitive and what is permanent, what is prescientific and what is supra-scientific in great myths."[50] While he discerns the inadequacy of purely rational approaches to the world, he does not relate the scientific to the religious insights. He simply wants

48. Langdon Gilkey, *Religion and the Scientific Future: Reflection on Myth, Science, and Theology* (New York: Harper & Row, 1970), 85.

49. Reinhold Niebuhr, *The Nature and Destiny of Man: A Christian Interpretation*, vol. 1, *Human Nature* (1941; reprint, New York: Scribner, 1964), 24.

50. Reinhold Niebuhr, "The Truth in Myths," in *Evolution and Religion*, 93.

to keep each of them in check so that they do not conflict with each other.

H. Richard Niebuhr, in his widely read book *Radical Monotheism and Western Civilization* (1960), has a long chapter, "Radical Faith and Western Science," discerning a parallel structure between the closed-society faith in religion and the closed-society faith in science. He is not worried that science would conflict with the religious element in religion but rather with the dogmatic truth systems of a closed-society faith. Niebuhr's argument could be interpreted to mean that belief in God the creator and sustainer of all things does not exclude the notion of evolution and indeed might even imply it. But this remains on the level of speculation. He does not mention evolution or its major interpreters. Neoorthodox theology was so intent on defining its own task of espousing God's word that it neglected the actual dialogue with other disciplines.

b. The Continental Way: From Apologetic Discernment to Withdrawal

Apologetic Discernment

The Lutheran theologian Christoph Ernst Luthardt (1823-1902), after 1856 professor of systematic theology and New Testament exegesis at Leipzig, Germany, emphasizes in his *Die christliche Glaubenslehre gemeinverständlich dargestellt (Exposition of the Christian Faith)* that the Bible is not a book about scientific research or scientific knowledge of nature, but a book of religion which has to do with humanity's relationship to God and with the relationships between people on earth.[51] Referring to the materialism debate which arose in 1852, Luthardt distinguishes between a psychological materialism that denies the existence of the soul and a later cosmological materialism which rejects the existence of the absolute Spirit and tries to reduce everything existing to matter.[52] This

---

51. Cf. Christian Ernst Luthardt, *Die christliche Glaubenslehre gemeinverständlich dargestellt*, 2nd ed. (Leipzig: Dörffling & Franke, 1906), 212.

52. Cf. for the following, Christian Ernst Luthardt, *Die modernen Weltanschauungen und ihre praktischen Konsequenzen: Vorträge über Fragen der Gegenwart aus Kirche, Schule, Staat*

kind of materialistic thinking is for Luthardt a result of pantheism, because it explains being from the idea. The absolute idea is everything, it reproduces being and permeates it. The Spirit then posits matter. Luthardt rejects this Hegelian dialectic. He notes that if all life is the movement of matter and all development is the development of matter, then there is no higher Spirit possible which sets a goal and an aim for the developmental process. This means that theology would be obsolete; all that would be left is a causality with its cause-and-effect sequence.

This kind of pantheism actually becomes a monism and a materialistic pantheism. The diversity of the world is then nothing but the progressive development of the world's actual foundations and beginnings. Darwinism fits well into this kind of thought pattern. Yet Luthardt asks how progress and development are possible if the effect cannot contain more than is contained in its cause. The world cannot be its own creator. If one dissociates God from the material realm, there is no spiritual world possible, not even a humanity. Luthardt concludes: "The result of the development is either pessimism or Christendom."[53] Either the world makes no sense whatsoever and one must despair, or one assumes some kind of spiritual meaning or beliefs. Materialism too is a faith; it is not a fact.

The Lutheran theologian Otto Zöckler (1833-1906), after 1866 professor of historical and exegetical theology at the University of Greifswald, Germany, was less confrontative than Luthardt. He attempted to demonstrate that the principles of good science and good theology do not necessarily conflict. Victor Schultze rightly claims in his article on Zöckler that Zöckler's scholarship "was rated very high as was his authority as a theologian in the realm of natural science."[54] Zöckler, who taught in Greifswald most of his life, had indeed an astounding command of scientific knowledge especially in its historic dimension. In his two-volume *Geschichte der Beziehungen zwischen Theologie und Naturwissenschaft mit besondrer Rücksicht auf Schöpfungsgeschichte* (*The History of the Relationship between Theology and the Natural Sciences with Special Attention to the Creation*

---

und Gesellschaft im Winter 1880 zu Leipzig gehalten (Leipzig: Dörffling & Franke, 1891), 169f., in his lecture "Der Materialismus und seine Konsequenzen."

53. Luthardt, *Die modernen Weltanschauungen und ihre praktischen Konsequenzen*, 185.

54. Victor Schultze, "Zöckler, Otto," in *The New Schaff-Herzog Encyclopedia of Religious Knowledge*, 12:520.

Narrative), he shows his immense erudition in the history of science and his critical awareness. For instance, while he knows that Darwin rejects a Christian teleological worldview, he realizes that Darwin "does not regard the sequence of the main events in the life of nature and humanity as 'the result of blind accident.'"[55] Zöckler concludes that Darwin's teachings do not contain anything which would necessitate the abandonment of the Christian theistic notion of creation (719). While Zöckler rejects Darwinism as a pathological disease which eventually will run its course, he is convinced that true to the Pauline saying that "all things are yours" (1 Cor. 3:21), the theological doctrines of creation and providence and the understanding of humanity's original state can gain new insights from the findings of Darwin (798ff.). Zöckler introduces here the concept of a theory of concordance through which the findings of evolutionary speculation insofar as they are scientifically proven complement the assertions of theology.

The hypothesis of concordance or of harmonizing is also employed in his lectures, *Die Urgeschichte der Erde und des Menschen (The Primal History of the Earth and Humanity)*. He claims that the contempt with which some scientists treat the first chapters of the Bible can only be the result of misinformation about the biblical Christian worldview as it pertains to the creation of the universe.[56] The eternal and infinitely grand perception contained in these stories has room for all the scientific details through which sober empirical research will enrich our understanding of how creation has occurred and still does occur. Even concerning the six days of creation there is a concordance between the record of geology and the book of Genesis (42). This mutually complementing avenue had already been pursued by George de Cuvier at the beginning of the nineteenth century, and according to Zöckler there seems to appear "an ever stronger consensus of all scientists even in Germany in this area which will soon lead to a complete victory over any contrary perspective" (48f.).

Zöckler's optimism is based on his historical research. He realized

---

55. Otto Zöckler, *Geschichte der Beziehungen zwischen Theologie und Naturwissenschaft mit besondrer Rücksicht auf Schöpfungsgeschichte*, vol. 2 (Gütersloh: C. Bertelsmann, 1879), 642f. Parenthetical references in the rest of this paragraph are to pages in this work.

56. Otto Zöckler, *Die Urgeschichte der Erde und des Menschen: Vorträge gehalten zu Hamburg im März 1868* (Gütersloh: C. Bertelsmann, 1868), 1f. Parenthetical references in the rest of this paragraph are to pages in this work.

that the claim made by the "modern fanatics of unbelief" that natural science sooner or later will do away with religion and with the Christian faith is simply not true.[57] There is no correlation between a comprehensive scientific education and religious unbelief. In each epoch there have been conservative and decidedly irreligious scientists, but most have pursued a middle course. He is convinced that the future does not belong to materialism but to a true empiricism that collects and analyzes the experiences available within the realm of the visible. "True scientists will time and again be able to read from the two texts placed alongside each other, from the Book of Nature and from the Book of Revelation. They will over and over again return to the religion of Kepler and Galilei, of Haller and Euler, and of Cuvier and Agassiz."[58] Of course, Zöckler concedes that in the future some will also radically deny the existence of everything which is not visible and tangible. But the true representatives of science will overcome these destructive forces.

True witnesses to God in the realm of nature will never die out as long as nature remains, because it is God's. Therefore human witness to the divine truth and grandeur contained in it will never be wanting. That nature witnesses to God is especially emphasized in Zöckler's earlier publication, *Theologia naturalis (Natural Theology)*. According to the maxim *credo ut intelligam* (I believe in order to understand), he wants to "explain, complete, and confirm the immediate revelation of God through the mediated one which is given in nature."[59] The book of nature will illustrate the book of the Bible, while the latter will explain the former. According to Zöckler, such a positive theology of nature will expand and illustrate the organic development of dogmatics. How nature and Bible come together can be seen especially well in the metaphors and parables of the Old and New Testaments.[60] The biblical symbols relating nature as well as its picturesque language exemplify the illus-

57. Otto Zöckler, *Gottes Zeugen im Reich der Natur: Biographien und Bekenntnisse aus alter und neuer Zeit,* 2nd ed. (Gütersloh: C. Bertelsmann, 1906), 482.

58. Zöckler, *Gottes Zeugen im Reich der Natur,* 485.

59. Otto Zöckler, *Theologia naturalis: Entwurf einer systematischen Naturtheologie vom offenbarungsgläubigen Standpunkte aus,* vol. 1, *Die Prolegomena und die specielle Theologie enthaltend* (Frankfurt am Main and Erlangen: Heyder & Zimmer, 1860), 6. Vol. 2 never appeared. This may perhaps serve as an indication that his project was more difficult than he had initially envisioned.

60. Zöckler, *Die Prolegomena und die specielle Theologie enthaltend,* 200f.

trative character of nature for God's revelation. Theological insight and scientific research do not go in separate or opposite ways but complement and in some respects even correct each other. Revelation and God's action occur not in a realm removed from the natural world but in the midst of nature. Nature is fundamentally the arena and medium of God's action. Otto Zöckler impresses with both his historical and scientific erudition and the ease with which scientific knowledge, for him, complements theological insights. He might be overly optimistic in asserting the complementarity of scientific knowledge with theological insights. However, at least he sees a vigorous engagement with the sciences as indispensable for theological assertions. With this approach he differed from most of his contemporaries.

Albrecht Ritschl and His Followers

Albrecht Ritschl first taught theology in Bonn and then in Göttingen. He was influenced by Kantian ethics, Schleiermacher's theology, and the legacy of the Lutheran Reformation. As one of the leading figures of late nineteenth-century theology in Germany, we discern in him a nearly complete withdrawal from nature. We would not expect that Ritschl would deal extensively with the topic of creation in his epoch-making three volume *The Christian Doctrine of Justification and Reconciliation.* Yet notably he declares that "theology has to do, not with natural objects, but with states and movements of man's spiritual life."[61] Certainly one can agree that theology is an intellectual enterprise and is concerned primarily with the spiritual side of humanity. Yet human nature consists of body and spirit, and the spiritual certainly affects the corporeal as Ritschl attests when he talks about the manifestation of human sinfulness. But Ritschl does not want to engage in a dialogue with the sciences or with the materialistic mind-set in the second part of the nineteenth century.

In his *Instruction in the Christian Religion,* in which we would expect a survey of the Christian faith, again the topic of creation is completely

---

61. Albrecht Ritschl, *The Christian Doctrine of Sanctification and Reconciliation: The Positive Development of the Doctrine,* trans. A. B. Macaulay et al. (New York: Scribner, 1900), 20.

missing. Only in a short sentence does Ritschl explain that God is the creator of the universe whose will determines everything toward God.[62] In a footnote he explains that "the conception of the creation of the world by God is entirely outside of all observation and ordinary experience, and therefore outside of the realm of scientific knowledge, which is limited by these." Therefore one cannot talk in analogy to natural causes about God creating the world. The origin and basis of the Christian faith and of science are different from one another. The Christian faith originates in a special revelation, and science is based on experience. Ritschl sees no point in engaging in a dialogue with science. For this reason materialistic claims cannot touch his theological assertions since they stem from a different source. Ritschl had not yet realized — or did not want to realize — that the scientific conquest of everything natural eventually will leave no room for a separate spiritual category. Either theology aligns itself with science or it simply dies of atrophy. Yet Ritschl's theocentric approach made theology attractive for theologians, since they could avoid dealing with the all-embracing scientific mind-set. This demarcation between the spiritual and the worldly, aided by a heavy dose of Kantian philosophy, is especially noticeable in Kaftan's theology.

Julius Wilhelm Martin Kaftan (1848-1926), professor first in Basel and later in Berlin, was heavily influenced by Ritschl but was not an actual pupil of his. According to Kaftan, the Christian worldview must correspond to the Christian faith in God.[63] The assertion of God as the almighty Lord of the universe is only another expression for the faith which acknowledges that the world is completely dependent on God. If God's almighty actions are guided and motivated by holy love, then everything in the world is ordered according to that love.

Scientific knowledge cannot endanger these theological statements since it does not touch the ultimate and highest questions. There are no ultimate causes for scientific knowledge possible. Our knowledge of nature is mediated through the senses. Their experience extends in space and time and can be subjected to experimental verification. This pro-

---

62. Albrecht Ritschl, *Instruction in the Christian Religion* (§11), in Albert T. Swing, *The Theology of Albrecht Ritschl*, together with *Instruction in the Christian Religion*, trans. Alice M. Swing (London: Longmans, Green, 1901), 183.

63. Cf. for the following, Julius Kaftan, *Das Wesen der christlichen Religion* (Basel: Bahnmaier's Verlag, 1881), 392f.

vides science with a strong objectivity which, however, rests in the phenomenal. "No created spirit penetrates into the interior of nature."[64] Therefore this objectivity remains relative; it is knowledge from outside.

By distinguishing between the phenomenal and the noumenal (as Kant does), Kaftan can discern different ways in which knowledge occurs in various fields. But this knowledge cannot be made uniform. Therefore both idealism, which reduces nature to spiritual processes, and materialism, which denies the independence of the spiritual life, transgress their proper boundaries. Similarly, when today more and more details become known about the process of evolution, this cannot diminish belief in creation. Bringing together the biblical creation narratives with the natural sciences, as is done in one strand of apologetics, makes no sense. For faith the thought of creation as such is of primary importance while the details of the arrangement of nature and the process of evolution are secondary.[65]

The worldview of science which perceives the world of phenomena is only a construct of the thinking spirit to master the real world. This worldview, which initially looks materialistic, is ultimately an idealistic construct, a means for our mind to perceive the world. Yet the same order that we recognize in science we also accept in faith as an expression of God's will.[66] The ultimately real for Kaftan is not the phenomenal world but the noumenal which is accessible through the immediate encounter with God. Therefore the world is given a certain independence, yet one that does not intersect with God's own sphere. God therefore is divorced from the world, and nature has no intrinsic value.

We conclude our brief survey with Johann Wilhelm Herrmann (1846-1922), professor in Marburg and influential for Karl Barth (1886-1968) and Rudolf Bultmann (1884-1976), who continued Kaftan's bifurcation of God and nature. Herrmann rejected traditional apologetics which attempt to establish the legitimacy of religion through science.[67] Such an endeavor would bring religion so far into the realm of science

---

64. Julius Kaftan, *Dogmatik* (Freiburg: J. C. B. Mohr, 1897), 104.

65. Cf. Kaftan, *Dogmatik*, 227.

66. Cf. Kaftan, *Dogmatik*, 250f.

67. For the following, cf. Wilhelm Herrmann, "Die Lage und Aufgabe der evangelischen Dogmatik in der Gegenwart" (1907), in Wilhelm Herrmann, *Schriften zur Grundlegung der Theologie*, pt. 2, ed. Peter Fischer-Appelt (Munich: Chr. Kaiser, 1967), 52ff.

that it would be suffocated. He wants to leave science to its inexhaustible task and philosophy to its task of shaping reality through reason. He would even agree that a true scientist and a truly pious person will reject a monism which thinks there is only one reality. Yet next to scientific knowledge there are also the forces of the subjective element which produce a liberating energy and betray a different understanding of the natural order. Though everybody lives in the real, tangible, and empirical world, there is an energy of life in the human individual that brings forth spiritual creativity and must be acknowledged as an independent phenomenon in every human being.

Like Friedrich Daniel Schleiermacher (1768-1834), Herrmann introduces a subjective element he calls religion, but does not want to restrict it to pure subjectivity. His approach becomes most evident when he states: "All revelation is self-revelation of God."[68] Since God is almighty, we cannot find God on our own but God will be found when God touches us in such an irresistible manner that we subject ourselves completely to God.

It may seem strange at first to discover that Herrmann ardently defends miracles. He maintains that "it is a matter of life and death for religion whether it can still assert the distinction of the miracle from a naturally understood event."[69] The reason for his stand is that on the one hand he views the laws of nature as an orderly context which is presupposed in our thinking as that which allows us to experience space and time. In such a context there is no miracle. Yet next to this experiential world there is also the world of faith which perceives the loving care of God's power. When this understanding becomes powerful in us, the other recedes. And when the experiential dominates again, faith takes a back place. Herrmann even sees a source of our energy in the fact that depending on the causal context, we oscillate between the thought of divine care and the thought of nature.

There is an inner tension in us, something irrational without which we cannot even think of life. God's activity, which is basically miraculous, occurs *supra et contra naturam,* beyond and against nature.[70] Such

---

68. Wilhelm Herrmann, *Offenbarung und Wunder* (Gießen: Alfred Töpelmann, 1908), 10.

69. Herrmann, *Offenbarung und Wunder,* 35.

70. Herrmann, *Offenbarung und Wunder,* 70f.

faith in God which draws us into itself cannot be transformed into some kind of commonly valid knowledge. It is derived from the existential experience of the individual. Herrmann therefore can assert faith as the most plausible entity in human life and at the same time affirm the natural context, including its materialistically sounding presuppositions. Herrmann would not surrender to materialism because it disclaims everything beyond its material basis. Nonetheless, he could understand materialistic disclaimers, since he knew that faith cannot be demonstrated but is strictly private business.

This individualistic privatization of faith which shaped much of twentieth-century theology failed to assert faith alongside a scientific and sometimes materialistic understanding of nature. It facilitated a retreat to the interior side of humanity, to that which is subjective, nondemonstrable, and, in the eyes of its critics, mushy and wishful thinking. Here the minority report of Zöckler and others demands more respect. Though it may have been overly optimistic in tracing the divine in nature, it was convinced that faith was not only compatible with science but that science, unless it was reduced to ideology, could not even infringe upon the religious dimension of nature. Since that kind of apologetics started with the presupposition that the world was God's world, it was not afraid of the scientific discovery of the world and its intricate interrelatedness. It did not need to retreat to the interior side of humanity, though it was aware that this side existed. Perhaps it was because of the misunderstood Kantian attempt to provide room for faith that religion became relegated to the supernatural and therefore materialism could dominate the scene as seen in Ernst Haeckel's monistic worldview.

When we consider for a moment the Reformation emphasis that God is the creator, sustainer, and redeemer of humanity and the world around us, the second part of the nineteenth century shows a very different picture in the heartland of the Reformation. Apart from a few apologetic endeavors, most prominently that of Otto Zöckler, the world comes hardly into focus. God only touches the interior side of humanity. This retreat from the external and tangible was facilitated by the Kantian distinction between the phenomenal and the noumenal, and also by Schleiermacher's claim that religion concerns itself primarily with feeling and intuition. Beyond that, the materialistic claim that everything existing can be reduced to matter and is subjected to an all-embracing cause-and-effect system left, in the eyes of most theologians, no other choice than to escape to something beyond the created order. Yet the scientific outlook changed drastically at the turn of the century. Since, however, this emphasis of the flight from creation was so pervasive in theology, even the new voices such as Karl Barth, Friedrich Gogarten (1887-1967), and Rudolf Bultmann continued this kind of escapist tendency with the consequence that theology turned itself more and more into a ghetto enterprise abandoning the dialogue with the other disciplines.

If it is possible to learn from history, we should affirm, along with the theologians just surveyed, that the Christian faith indeed is not an epiphenomenon of matter as materialists claimed, but an integrative force allowing for an expanding vision beyond the material base. Yet we should be extremely hesitant to relegate faith completely in the otherworldly or the supranatural since it would then be extremely difficult to exert any guiding vision on the material world and those who inhabit it unless one turns faith into ideology.

c. The Beginning Dialogue with the Natural Sciences in the Twentieth Century

In the USA and in Great Britain the dialogue with the natural sciences has enjoyed a long and uninterrupted tradition. This may in part be due to the fact that in the nineteenth century Great Britain was leading in scientific discoveries and in its wake America put these discoveries to technological use. More importantly, while on the Continent the religious picture was still homogeneous, in Great Britain there had emerged many independent religious groups, such as the Puritans, the Dissenters, the Baptists, the Methodists, and numerous others. North America even implemented the separation of church and state, which meant that no religious group was supposed to be part of the establishment. Yet in Germany and in most other countries on the Continent with the exception of France, the princes and kings were still the religious overlords of their people. They controlled theological education and what was to be taught and preached. In the USA, however, the founding figures, such as Thomas Jefferson (1743-1826), George Washington (1732-99), and Benjamin Franklin (1706-90), were free thinkers with strong Deistic inclinations, and in the nineteenth century the Unitarians on the East Coast had a decided influence on the religious and cultural life of many people.

Since the establishment of the "new world" in North America carried with it decidedly religious overtones, it was clear that even if religion was treated as a natural phenomenon, the discoveries in nature should enhance this religious characteristic and not diminish it. Indicative for this mood, for instance, are the McNair Lectures at the University of North Carolina. They should show "the mutual bearing of science and theology upon each other," and should "prove the existence and attributes, as far as may be, of God from nature."[71] In Great Britain also, natural theology and the defense of a reasonable belief in God has a long tradition starting with the Deists and free thinkers of the eighteenth century. The Gifford Lectures in Scotland are dedicated to the area of natural theology, and the Bampton Lectures in Oxford serve a general apologetic use. In Germany such endowed lecture series are unknown, and even the ecclesial academies in which many current issues are discussed were only started after World War II. Except for a few solitary individuals who vigorously engaged themselves in the dialogue between theology and science such as Karl Heim (1874-1958), who for most of his academic life was professor of systematic theology in Tübingen, and Arthur Titius (1864-1936), professor of theology in Berlin, there were only occasional comments by scientists, such as Albert Einstein, Max Planck, and Werner Heisenberg. Especially neo-Reformation theology, with its insistence on the primacy of God's word, proclaimed a timeless gospel instead of engaging itself in a dialogue with the natural sciences.

How little theology had to contribute can be seen with **Günter Howe** (1908-68).[72] He had obtained his Ph.D. in mathematics and met Karl Barth, whose theology had opened new possibilities for him. A dramatic personal discovery, a quantum jump of history occurred for him when he noticed Barth's concern for the laity and their responsibilities as well as his emphasis on the triune God as the starting point of theol-

71. Cf. C. A. Coulson, *Science and Christian Belief* (Chapel Hill: University of North Carolina Press, 1955), in his preface.

72. Cf. Guy Clicqué, *Differenz und Parallelität. Zum Verständnis des Zusammenhangs von Theologie und Naturwissenschaft am Beispiel der Überlegungen Günter Howes* (Frankfurt am Main: Peter Lang, 2001).

ogy. Howe discovered analogous structures between Barth's doctrine of God and Niels Bohr's theory of complementarity.[73] Encouraged by the physicist and philosopher Carl Friedrich von Weizsäcker (b. 1912), he thought a dialogue on these similarities would be fruitful. As a result of his enthusiasm, approximately twenty-five theologians, physicists, mathematicians, and chemists convened in Göttingen in 1949 for a conference. This conference led to annual meetings (which later also included philosophers) and lasted till 1961. Then they were institutionalized in the Protestant Institute for Interdisciplinary Research (Forschungsstätte der Evangelischen Studiengemeinschaft: FEST) in Heidelberg. This institute is supported by the Protestant Churches in Germany and the Protestant academies.

Barth, who had been invited to the first conference, excused himself with the remark that he was too busy, and Heim, who would have liked to come, was at that time too sick to travel. Barth regarded such a dialogue as a waste of time, because he thought it would only resurrect the old natural theology. Therefore he could have contributed only in a negative way and polemicized against the scientists who did participate.

In 1958, under the direction of Willem Visser 't Hooft (1900-1985), the secretary-general of the World Council of Churches, an International European Conference was convened in Bossey, Switzerland, at which the members of the Göttingen circle participated. On account of the development of atomic weapons, the dialogue was now directed toward a responsible natural science and a responsible society. With his increasing interest in a "responsible natural science," Howe struck up a friendship with Heinz-Eduard Tödt (1918-91), professor of systematic theology at Heidelberg University. They conducted lecture series and

---

73. Cf. Günter Howe, *Die Christenheit im Atomzeitalter: Vorträge und Studien* (Stuttgart: Ernst Klett, 1970), and esp. his contribution of 1958, "Das Göttinger Gespräch zwischen Physikern und Theologen" (The Göttingen discussion between physicists and theologians) as well as the epilogue by Hermann Timm, "About Günter Howe's Theological Journey," and the foreword by Carl Friedrich von Weizsäcker. Cf. also Günter Howe and Heinz Eduard Tödt, *Frieden im wissenschaftlich-technischen Zeitalter: Ökumenische Theologie und Zivilisation* (Peace in the scientific technical age: Ecumenical theology and civilization) (Stuttgart: Kreuz-Verlag, 1966) and Günter Howe, *Gott und Technik: Die Verantwortung der Christenheit für die wissenschaftlich-technische Welt* (God and technology: Christianity's responsibility for the scientific-technical world), ed. Hermann Timm, introduction by Heinz Eduard Tödt (Hamburg: Furche, 1971).

seminars together, and Howe was named professor at the theological faculty in Heidelberg. He was a leader in the ecumenical peace discussions until he suddenly died, barely sixty years old, in 1968.

Another natural scientist, **Pascual Jordan** (1902-80), professor of physics at Hamburg University, did not directly enter the theological arena. Yet he entered the dialogue in 1963 with the publication of *The Natural Scientist Confronting the Religious Issue (Der Naturwissenschaftler vor der religiösen Frage)*.[74] He convincingly argued that the barriers to religion no longer existed. These barriers, which were erected in the nineteenth century, were no longer obstacles between the disciplines. How large the interest was in such a dialogue is indicated by the fact that his book had a second printing within a year of the first.

That the dialogue gained momentum can perhaps best be seen in the collaborative work of **A. M. Klaus Müller** (1931-95) with Wolfhart Pannenberg (b. 1928). Müller became known among theologians after their joint publication *Towards a Theology of Nature (Erwägungen zu einer Theologie der Natur)*.[75] This publication resulted from the Karlsruhe Conversations of Physicists and Theologians of the Younger Generation (Karlsruher Physiker-Theologen-Gespräche der jüngeren Generation), a sequel to the Göttingen Conferences which had been initiated by Günter Howe. Müller contributed to this publication an essay "On the Philosophical Treatment of Exact Research and Its Necessity." According to Müller, a physicist "knows that he does physics, and whether and when he does a good job at it, but he is not sure what physics is."[76] To determine this the scientist must compare physics with something else. The philosopher who thinks about physics searches for a horizon within which the truth of physics becomes visible. From this it follows that philosophical questions open more clearly the horizon of consciousness for those who conduct research in the natural sciences. These philosophical questions lead to new views and point to new areas of research. Müller illustrates this by using examples from the history of the natural

74. Pascual Jordan, *Der Naturwissenschaftler vor der religiösen Frage: Abbruch einer Mauer* (Oldenburg: Gerhard Stalling, 1963), esp. 357.

75. A. M. Klaus Müller and Wolfhart Pannenberg, *Erwägungen zu einer Theologie der Natur* (Gütersloh: Gerd Mohn, 1970).

76. Cf. for the following, A. M. Klaus Müller, "Über philosophischen Umgang mit exakter Forschung und seine Notwendigkeit," in *Erwägungen zu einer Theologie der Natur*, by Müller and Pannenberg, 9.

sciences. Furthermore, he emphasizes that a critical examination of science's own presuppositions often leads to progress in the field. Therefore dialogue is necessary to clarify one's own thinking and to push it ahead.

Müller is a representative of the present generation who is concerned about a dialogue. Though these scholars are trained in classical issues, they are concerned about the present situation. This means that the dialogue is largely determined by the environment in which it takes place. This is also true on the theological side. Here it was the French Jesuit and paleontologist Pierre Teilhard de Chardin (1881-1955) who initially drew the attention of theologians to the natural sciences and who guided them to a cosmic understanding of the world. The Roman Catholic theologian Karl Schmitz-Moormann (1928-96) from Dortmund University has the distinction of having translated Teilhard's works into German. Sigurd Daecke (b. 1932) investigated the significance of Teilhard for Protestant theology in his theological dissertation: *Teilhard de Chardin and Protestant Theology: The Worldliness of God and the Worldliness of the World.*[77] He did not merely portray Teilhard's theology but also asked how prominent Protestant theologians of the twentieth century such as Paul Tillich, Friedrich Gogarten, Dietrich Bonhoeffer, Gerhard Ebeling (1912-2001), Wolfhart Pannenberg, and Karl Heim understood the relationship between God and the world. Daecke pointed out that both Heim and Teilhard have a common concern but their work led them to opposite results.

One year prior to Daecke, Jürgen Hübner (b. 1932), in his dissertation *Theology and the Biological Doctrine of Evolution,* had pointed to the most important theological efforts in Germany since Darwin to interpret the theory of evolution.[78] Already in 1965 Günter Altner was concerned with similar matters, as the title of his dissertation reveals: *Creation and Evolution in Protestant Theology from Ernst Haeckel to Teilhard de Chardin.*[79] Altner (b. 1936) is especially well equipped for this dialogue, since he has doctorates in both theology and biology.

77. Sigurd Daecke, *Teilhard de Chardin und die evangelische Theologie: Die Weltlichkeit Gottes und die Weltlichkeit der Welt* (Göttingen: Vandenhoeck & Ruprecht, 1967).

78. Jürgen Hübner, *Theologie und die biologische Entwicklungslehre: Ein Beitrag zum Gespräch zwischen Theologie und Naturwissenschaft* (Munich: C. H. Beck, 1966).

79. Günter Altner, *Schöpfungsglaube und Entwicklungslehre in der protestantischen Theologie zwischen Ernst Haeckel und Teilhard de Chardin* (Zürich: Evang. Verlagsanstalt, 1965).

That the dialogue between theology and the natural sciences has gained momentum is shown by the First European Conference on Science and Religion, which took place March 13-16, 1986, in the Protestant Academy in Loccum, Germany. The conference was organized on the topic "Evolution and Creation." Prominent scientists and theologians such as Manfred Eigen (b. 1927), Arthur Peacocke (b. 1924), and Jürgen Hübner participated in that conference. By now this conference, which is conducted in English, has become a biannual event and takes place in different European countries (Germany, the Netherlands, Switzerland, Italy, Poland, etc.). Its results are published in the yearbook of the European Society for the Study of Science and Theology (ESSSAT), the parent organization for these conferences, which was established in 1990 "in response to the growing sense of a change of climate in the relationship between science and theology."[80] ESSSAT is an international European organization which attempts to foster dialogue and cooperation between academic scientists and theologians.

Among organizations of longer standing, we must mention once more the Forschungsstätte der Evangelischen Studiengemeinschaft (FEST) in Heidelberg (Protestant Institute for Interdisciplinary Research), which enjoys the official support of the Protestant Churches in Germany. It supports research, consultations, and publications that deal with issues arising from the interface between church and society, including such areas as the social and natural sciences, law, ecology, and politics. Günter Altner, Jürgen Hübner, Jürgen Moltmann (b. 1926), Horst W. Beck (b. 1933), and many others have been active in its projects. FEST has also sponsored a major contribution to the dialogue between theology and the natural sciences with the publication of *The Dialogue between Theology and the Natural Sciences: A Bibliographic Report,* which was edited by Hübner.[81]

The Deutsches Institut für Bildung und Wissenschaft (German Institute for Education and Science) and its special department, Institut für Wissenschaftstheoretische Grundlagen-forschung (Institute for Investigation into the Foundations of Theoretical Knowledge), must also

---

80. *Who Is Who in Theology and Science: 1996 Edition,* comp. and ed. the John Templeton Foundation (New York: Continuum, 1996), 580.

81. Jürgen Hübner, *Der Dialog zwischen Theologie und Naturwissenschaft: Ein bibliographischer Bericht* (Munich: Christian Kaiser, 1987).

be mentioned. One of its major projects was "to work out the intellectual and spiritual trends of the present age and to analyze them as a challenge of belief to one-dimensional rationality."[82] This project resulted in the publication of four volumes.[83] This institute was founded in 1970 and is largely supported by the Roman Catholic Church. It concerns itself with interdisciplinary issues and its research is done on a voluntary basis by some one hundred Protestant and Roman Catholic university professors from various disciplines. The founding director of this institute is Hugo Staudinger (b. 1921), a retired professor at Paderborn University.

The Karl Heim Gesellschaft (Karl Heim Society) is another organization promoting the dialogue between science and theology. This society of approximately eighty members and 550 supporting friends preserves the legacy of Karl Heim and attempts to provide a Christian orientation within a scientific-technological world through publications, seminars, and lectures. It was founded in 1974 and publishes a journal, *Evangelium und Wissenschaft (Gospel and Science)*. Since 1988 its president, Hans Schwarz (b. 1939), edits its yearbook, *Glaube und Denken (Faith and Thought)*. The journal contains essays, information about conferences, and book reviews, while the yearbook presents invited papers and usually a reprint of a smaller piece by Karl Heim. The Karl Heim Society also maintains the Karl Heim Archives, which are housed in the Bengel House in Tübingen, a study center and boarding house associated with the university. Through seminars and annual meetings this society fosters the dialogue between theology and science. The society is also supported by the Protestant Church in Württemberg.

In 1981 Horst W. Beck (b. 1933), a scientist and theologian, left the Karl Heim Society, taking with him his own circle of friends and supporters. Through seminars and conferences he has attempted to offer a

---

82. *Who Is Who*, 587.

83. Vol. 1: Hugo Staudinger and Wolfgang Behler, *Chance und Risiko der Gegenwart: Eine kritische Analyse der wissenschaftlich-technischen Welt* (Paderborn: Ferdinand Schöningh, 1976); vol. 2: Hugo Staudinger and Johannes Schlüter, *Wer ist der Mensch? Entwurf einer offenen und imperativen Anthroplogie*, 2nd ed. (Stuttgart: Burg, 1991); vol. 3: Ludwig Kerstiens, *Verbindliche Perspektiven menschlichen Handelns: Zum Problem der Gültigkeit, Anerkennung und Vermittlung von Werten und Normen* (Stuttgart: Burg, 1983); vol. 4: Hugo Staudinger and Johannes Schlüter, *Die Glaubwürdigkeit der Offenbarung und die Krise der modernen Welt —
Überlegungen zu einer trinitarischen Metaphysik* (Stuttgart: Burg, 1987).

conservative alternative to the Karl Heim Society. He publishes *Wort und Wissen (Word and Knowledge: Impulses, Materials, Courses for Christian Alternatives in Science, Technology, and Society)*. This endeavor is supported by the Studiengemeinschaft Wort und Wissen (Research Society: Word and Knowledge), and its publications are issued by the evangelical publishing house Hänssler in Stuttgart.[84] The main tendency is antievolutionary and creationistic. But Beck and his friends do not avoid contact with other-minded people. This is demonstrated, for instance, by a conference in 1984 at the Protestant Academy in Hofgeismar. Beck, Hübner, and Günter Ewald (b. 1929), a mathematician who has been closely connected with the Karl Heim Society and FEST, presented papers on the topic "Evolution and Creation." In contrast to the USA, the German ecclesial situation allows various groups to meet in dialogue about theology and the sciences without retreating into a more comfortable ghetto mentality.

Beyond the Continent, Christians in Science, formerly the Research Scientist's Christian Fellowship, in England must be mentioned. This group presently has a membership of approximately 750, composed of scientists, philosophers, theologians, ministers, and others with an interest in science and religion questions. This society attempts to promote a positive Christian view of nature, and discern the scope and limitations

84. Some of the publications of *Wort and Wissen* are: vol. 1: Horst W. Beck, *Biologie und Weltanschauung — Gott der Schöpfer und Vollender und die Evolutionskonzepte des Menschen* (Stuttgart: Hänssler, 1979); vol. 2: Joachim Scheven, *Daten zur Evolutionslehre im Biologieunterricht — Kritische Bilddokumentation* (Stuttgart: Hänssler, 1979); vol. 3: Dieter Bierlein, *Entscheidung und Verantwortung in kybernetischer Sicht* (Stuttgart: Hänssler, 1979); vol. 5: Werner Gitt, *Logos oder Chaos — Aussagen und Einwände zur Evolutionslehre sowie eine tragfähige Alternative* (Stuttgart: Hänssler, 1980); vol. 6/1: Horst W. Beck, *Schritte über Grenzen zwischen Technik und Theologie: Der Mensch im System — Perspektiven einer kybernetischen Kultur* (Stuttgart: Hänssler, 1979); vol. 6/2: Horst W. Beck, *Schritte über Grenzen zwischen Technik und Technologie; Schöpfung und Vollendung — Perspektiven einer Theologie der Natur* (Stuttgart: Hänssler, 1979); vol. 8: Horst W. Beck, Heiko Hörnicke, and Hermann Schneider, *Die Debatte um Bibel und Wissenschaft in Amerika — Begegnungen und Eindrücke von San Diego bis Vancouver* (Stuttgart: Hänssler, 1980); vol. 9: Edith Düsing and Horst W. Beck, *Menschenwürde und Emanzipation — Entfremdung und Konzepte ihrer Aufhebung — kritischer Traktat* (Stuttgart: Hänssler, 1981); vol. 10: Alma von Stockhausen, *Mythos — Logos — Evolution: Dialektische Verknüpfung von Geist und Materie* (Stuttgart: Hänssler, 1981); and Horst W. Beck, *Universalität und Wissenschaft: Grundriß interdisziplinärer Theologie* (Stuttgart: Hänssler, 1987; 2nd ed.: Weilheim-Bierbronnen: Gustav-Siewerth-Akademie, 1994).

of science in the modern world. It was founded in 1942, and since 1989 it has published a journal, *Science and Christian Belief.* It also stages an annual conference in London and sets up various study groups on topics such as the environment, miracles, and the use of minerals, and occasionally holds other meetings and conferences.

Significant also is the group Christians in Science Education, mainly a British organization set up to help people involved or interested in science education from a Christian viewpoint. It was established in 1989. In 1972 the Science and Religion Forum was founded to enable conversations between scientists, theologians, and clergy who wish to relate their scientific knowledge and methods of study to religious faith and practice. Active in this forum are the former archbishop of York, John S. Habgood (b. 1927), Arthur Peacocke from Oxford University, and Russell Stannard (b. 1931) of the Open University. The forum organizes an annual conference, and the papers of these conferences are usually published. Finally, one ought to mention the Society of Ordained Scientists, a fellowship of scientists within the ordained ministry of the Anglican Communion who meet for support and the furtherance of service in their scientific and spiritual endeavor. This fellowship was founded in 1987 and has an annual meeting and a retreat. It is also open to ordained members of other churches.

When we turn our attention to North America, we could mention a host of scholars. Most interesting are those theologians or scientists who have a second field of expertise. For instance, **Ian G. Barbour** (b. 1923), the grand seigneur of the dialogue between theology and science and a recipient of the Templeton Prize for Progress in Religion, graduated with a bachelor of divinity degree from Yale Divinity School and received a doctor of philosophy in physics from the University of Chicago. For a long time he was head of the department of religion and professor of physics at Carleton College in Northfield, Minnesota. He has a magisterial command of the dialogue between theology and the sciences which is most ably demonstrated in his seminal book *Issues in Science and Religion* and in his Gifford Lectures.[85] Mention must also be

85. Ian G. Barbour, *Issues in Science and Religion* (Englewood Cliffs, N.J.: Prentice-Hall, 1966); Barbour, *Religion in an Age of Science: The Gifford Lectures, 1989-1991*, vol. 1 (San Francisco: Harper, 1990); Barbour, *Ethics in an Age of Technology: The Gifford Lectures, 1989-1991*, vol. 2 (San Francisco: Harper, 1993).

made of William G. Pollard (1911-89), the first executive director at the Oak Ridge Institute of Nuclear Studies in Oak Ridge, Tennessee, and at the same time associate minister of an Episcopal congregation.[86]

Most interesting and important for the dialogue between theology and science, however, are the various organizations and centers of research. One of the oldest established organizations is the American Scientific Affiliation (ASA), which was founded in 1941 in Chicago. It is a fellowship of over two thousand Christians in the sciences organized in eighteen local sections in most of the fifty states of the USA, with additional members in forty-six other countries, committed to understanding the relationship of science to the Christian faith. The stated purpose of the ASA is "to investigate any area relating Christian faith and science" and "to make known the results of such investigations for comment and criticism by the Christian community and by the scientific community."[87] Full membership is restricted to persons who have at least a bachelor's degree in science and can assent to the affiliation's statement of faith, while associate membership is available to anyone interested in science who can assent to that statement. The statement of faith states that the Bible is the inspired word of God in matters of faith and conduct, that one recognizes his or her responsibility as a steward of creation to use science and technology for the good of humanity and for the whole world, and that God has created and does preserve the universe and has endowed it with a contingent order and intelligibility which can be scientifically investigated. This fellowship, which is evangelical in persuasion but advocates neither fundamentalism nor creationism, issues a quarterly journal, *Perspectives on Science and Christian Faith*, and the bimonthly *Newsletter* for its members. There are studies and symposia going on in a variety of areas such as biomedical ethics, creation, global resources, science education, etc. Divisions of that affiliation are the Affiliation of Christian Geologists and the Affiliation of Christian Biologists.

Evangelical in persuasion too is the AuSable Institute of Environmental Studies located in Mancelona, Michigan, and founded in 1980. It

86. William G. Pollard, *Chance and Providence: God's Action in a World Governed by Scientific Law* (New York: Scribner, 1958); and *Physicist and Christian: Dialogue between the Communities* (Greenwich, Conn.: Seabury, 1961).

87. *Who Is Who*, 566.

is a Christian environmental stewardship institute which sees its mission as bringing healing and wholeness to the biosphere and to the whole of creation. It offers programs and courses of study for college students, evangelical Christian colleges, denominations and churches, and the broader world community. Some ninety Christian colleges and seminaries in the United States and Canada participate in its programs.

Another organization of long esteem is the Institute on Religion in an Age of Science (IRAS), which had as its main promoter **Ralph Wendell Burhoe** (1911-97), director of the Center for Advanced Study in Religion and Science (CASIRAS) at the Unitarian Meadville/ Lombard Theological School in Chicago. It has 310 members, widely distributed geographically and with an international representation. Since 1954 it has staged every August the weeklong Star Island Conference near Portsmouth, New Hampshire. Scientists, theologians, and other interested experts treat a topic which arises from current scientific thought and fundamental religious issues. IRAS is an independent society of scientists, philosophers, theologians, and religious thinkers who attempt to understand the role of religion in a dynamic world investigated by science. It is an affiliate of the American Association for the Advancement of Science, and it organizes occasional conferences and meetings. It is a copublisher of the quarterly journal *Zygon: Journal of Religion and Science* and also publishes the *IRAS Newsletter,* which appears three times a year, and the *Science and Religion News,* published four times a year whose most recent issue was entitled *Science & Spirit.* While *Zygon* contains learned papers of noted scientists and theologians, the other two publications are mainly attempts to disseminate information. Especially in *Science and Religion News,* everything noteworthy appears that in the widest sense relates to the dialogue between theology and science, including a listing of journal articles, yearbooks, papers, and conferences.

A significant contribution is also made by the Chicago Center for Religion and Science, operated in collaboration with the Center for Advanced Study in Religion and Science and the Lutheran School of Theology at Chicago. This center was founded in 1988 and until his retirement in 2001 ably directed by **Philip Hefner** (b. 1932), professor of systematic theology at the Lutheran School of Theology. He is also currently the editor of *Zygon.* This shows the kind of interconnection that occurs in the various institutions and affiliations. The purpose of the Chicago Center is to provide a place of research and discussion between scientists, theologians, and other scholars on the most basic issues pertaining to the understanding of the world, the traditional faith, and how both can contribute to the welfare of the human community. The center supports the research of resident scholars, organizes conferences and seminars, and provides support for topical research workshops. Its newsletter, *Insights,* appears three times a year.

Another focal point for dialogue is the Center for Theology and the Natural Sciences (CTNS), founded and directed since 1981 by **Robert John Russell** (b. 1946). The center is affiliated with the Graduate Theological Union in Berkeley and has more than 500 members committed to the fruitful interaction of science and religion. It has been organized around three objectives: research, education, and public service. The CTNS program includes the following activities: bringing internationally distinguished senior scholars to its center for an extended period of research; staging international research conferences; offering courses for doctoral and seminary students; offering public lectures, forums, and other events for scientists, clergy, technologists, and the general public; engaging in certain research projects, such as the human genome project in its theological-ethical implications and a long-term joint research project with the Papal Observatory in Rome. Associated with this center are Ian Barbour, Nancey Murphy (b. 1951), and Ted Peters (b. 1941), among others. The center publishes the quarterly *CTNS Bulletin* with scholarly articles and book reviews and a monthly *Newsletter* with local and international news on conferences and programs.

Finally, for the USA, one should mention the Center of Theological Inquiry located on the campus of Princeton Theological Seminary in Princeton, New Jersey. The center is ecumenical, interreligious, and international in scope and brings together promising and outstanding scholars for conferences, consultations, and periods of residential scholarship. Special emphasis lies in interdisciplinary dialogue. An example of such dialogue is the activation of critical discussions for the purpose of renewing the responsiveness of theology to the issues of cosmology and ecology, social institutions, biology and personal ethics, culture and the world religions. Many well-known theologians and philosophers of religion have worked at this center which provides access to the libraries of Princeton Theological Seminary and Princeton University. The board of trustees of the center includes many well-known philanthropists.

In Canada mention must be made of the Pascal Centre for Advanced Studies in Faith and Science. It was founded in 1988 and is affiliated with Redeemer College at Ancaster, Ontario, with which it shares facilities. In 1990 the Pascal Centre began publishing its *Pascal Centre Notebook,* a quarterly newsletter. The center is Bible-oriented, and in the light of Scripture it attempts to fathom the relationships between theology, philosophy, and the natural sciences. To that end it features seminars each year on a variety of topics as well as conducting various research projects (for instance, in the history of science and hermeneutics by analyzing the interaction of cosmology and the Bible in modern Protestant Europe).

We conclude our brief survey by mentioning the Templeton Foundation, which in the last few years has focused more and more on the dialogue between theology and the natural sciences. This foundation, which is named after the Anglo-American investment banker **Sir John Marks Templeton** (b. 1912), who lives in the Bahamas and supports this foundation almost single-handedly, is located in Ipswich, Massachusetts. Since 1970 it has awarded the annual Templeton Prize for Progress in Religion, which is conferred at Westminster Abbey in London. This foundation has also published *Who's Who in Theology and Science,* which includes persons, organizations, and journals which further the dialogue between theology and the sciences. Furthermore, it issues the monthly periodical *Research News & Opportunities in Science and Theology,* gives awards to publications which further the dialogue between theology and the natural sciences, and promotes this dialogue by subsidizing seminars and lecture series in colleges and universities.

A further indication that the dialogue between theology and the sciences is advancing and doing well is the appearance of a new book series titled Theology and the Sciences. This series, which began in 1993 at Fortress Press in Minneapolis, Minnesota, is edited by Kevin J. Sharpe (b. 1950), who is also the editor of *Science & Spirit,* and includes among its advisers Ian Barbour, Philip Hefner, John Polkinghorne (b. 1930), and Robert John Russell. Authors in the series include Langdon Gilkey, Arthur Peacocke, and H. Paul Santmire (b. 1935).

d. Scientists at the Border (von Weizsäcker, Davies, Hawking, Tipler)

The biography of **Carl Friedrich von Weizsäcker** (b. 1912), the older brother of the former president of the Federal Republic of Germany, shows very clearly how this scientist has encountered the limits of what can be known and advocated by natural science. Already in his youth he was deeply impressed by Werner Heisenberg. Initially in his career he was a professor of theoretical physics at the universities of Strasbourg and, later, Göttingen. Then from 1957 to 1969 he was professor of philosophy at the University of Hamburg. For ten years he directed the Max Planck Institute for Research on the Living Conditions in a Scientific Technological World at Starnberg. Von Weizsäcker functions as a bridge builder between theology and the natural sciences.[88] He realized that the issues in modern nuclear physics had been fundamentally changed in 1945 by the first use of atomic bombs in Hiroshima and Nagasaki. Nuclear physicists could not return to their own field of inquiry as if nothing had happened. Now they were confronted with philosophical, ethical, and political questions. In this situation von Weizsäcker together with Günter Howe initiated the Göttingen colloquies between physicists and theologians (1949-61), which can be regarded as the starting point of the modern dialogue between the natural sciences and theology in Germany.[89]

In these conversations the participants attempted to fathom the limits of scientific technological knowledge and their responsibility toward the world. Theologians and scientists together should shoulder responsibility for the world by thinking through the consequences of the natural sciences. Such a feeling of responsibility paradigmatically shows that the old controversy between theology and the natural sciences can no longer continue. Now other issues which are more important and even life threatening need to be considered. While in the old conflict the question was raised whether God still had a place in a world which had become explicable by natural science and was largely

---

88. Cf. the comprehensive work of Deuk-Chil Kwon, *Carl Friedrich von Weizsäcker als Brückenbauer zwischen Theologie und Naturwissenschaft* (Frankfurt am Main: Peter Lang, 1995).

89. Cf. Harold P. Nebelsick, "Naturwissenschaft und die Zukunft der Theologie," *Glaube und Denken* 2 (1989): 49.

dominated by modern technology, now, according to von Weizsäcker, the question had to be turned around. It is no longer our question concerning God that needs to be discussed, but God's question concerning us, concerning the natural scientists and our responsibility for the world and our blame for its possible destruction.[90] Von Weizsäcker challenges natural scientists to take on ethical and political responsibility, since scientific investigation is no longer neutral in value when confronted with the issue of humanity's survival. It must now consider the ethical and political consequences of its findings. Therefore the ethical problems of scientific investigation occupy center stage in the dialogue between theology and science. The former governor of Saxony, Kurt H. Biedenkopf (b. 1930), is right in his assessment of von Weizsäcker: "As a person of both worlds, of the natural sciences and the humanities, he has embodied the dialogue between the natural sciences and the humanities."[91]

In von Weizsäcker's youth the dialogue between theology and the sciences was already taken for granted. He reminisces that on his twelfth birthday he was gazing at the stars above him on a pleasant summer's night: "In the inexpressible beauty of the starry sky God was somehow present. At the same time I realized that these stars were gaseous globes consisting of atoms and subject to the laws of physics. The tension between these two truths cannot be without resolution. But how can one solve it? Would it be possible to detect even in the laws of physics a reflection of the beauty of God?"[92] In addressing this question today, von Weizsäcker is guided by the scientist, theologian, and astronomer Johannes Kepler (1571-1630). The mathematical laws of nature disclose an unforeseen beauty: "The world is built according to the creative thoughts of God, this means in mathematical harmony. Humanity, created in God's image, is able to trace these thoughts. Natural science is divine service."[93] But scientific knowledge is also power even if it is not

---

90. Cf. also Carl Friedrich von Weizsäcker, *Deutlichkeit: Beiträge zu politischen und religiösen Gegenwartsfragen* (Munich and Vienna: Hanser, 1978), 155f.

91. Kurt H. Biedenkopf, "Das Recht der Utopie," in Carl Friedrich von Weizsäcker, *Das Ende der Geduld: Die Zeit drängt in der Diskussion* (Munich and Vienna: Hanser, 1987), 57.

92. Carl Friedrich von Weizsäcker, *Der Garten des Menschlichen: Beiträge zur geschichtlichen Anthropologie* (Munich: Carl Hanser, 1977), 553.

93. Von Weizsäcker, *Der Garten des Menschlichen*, 442.

sought for power's sake. Therefore there results an ethical perspective of the natural sciences.[94]

Since von Weizsäcker starts with the premise that God has created the world, it is theologically irrelevant for him to give the world a certain age, for instance 10 billion years, or "to assume that it has existed since an infinite time."[95] According to Augustine, creation is not an event in time but an act through which time itself is constituted. Therefore there exists no necessity why God should have preferred to create a finite instead of infinite time. The theory of whether an original big bang holds true or not is relatively unimportant for von Weizsäcker, since such scientific explanations always deal with something discovered within the world. In a similar way, von Weizsäcker faces the biological doctrine of organic development with relative ease. He asks himself: "Why should God have refrained from using the natural laws of growth and transformation in creating living beings?" (cf. 128). With a high degree of possibility, modern biology shows that life has developed from inorganic matter and that humanity, a relative latecomer on earth, has descended from higher animals. This method always seeks to trace back from what is at hand to something prior to it. In doing so "modern science excludes direct creation" (cf. 136). While this method does not allow us to prove the inorganic origin of life, or that humanity descended from animal-like forms of life, it would not make much sense to reject such hypotheses. Moreover, according to von Weizsäcker, these hypotheses do not touch the assertion of faith that God has created the world.

More problematic, however, is the fact that "faith in science plays the rôle of the dominating religion of our time" (so 12). The place once occupied by religion in the soul of most people is today occupied by science, or rather by faith in science and in the ability to shape the future. This is true even for people who understand themselves as skeptics or even enemies of science. "The romantic author who has written a book against the world view of science calls his publisher by telephone because he is late in his proofreading; and by this very act he tacitly bows before the god whom he defies in his writing" (14). Yet the possibility of

---

94. Cf. Carl Friedrich von Weizsäcker, *Wahrnehmung der Neuzeit* (Munich: Carl Hanser, 1983), 340.

95. Cf. Carl Friedrich von Weizsäcker, *The Relevance of Science: Creation and Cosmogony. Gifford Lectures, 1959-60* (New York: Harper & Row, 1964), 155. The parenthetical numbers in the following text refer to pages in this work.

dominating and modifying nature has only become a reality through the demythologization of the world and the radical monotheism introduced by the Judeo-Christian tradition. "Even modern secularization, even scientism cannot be understood" without this background (51). Von Weizsäcker does not only point to the biblical narrative which he thinks contains the possibility that creation can be handed over to humanity, but also to the fall, because actual world history only commences with the fall of Adam and Eve, a "history, which is a struggle between God and man from the outset. Man has deserted God and deserts him daily" (53).

Von Weizsäcker first attempted to describe scientifically what we can understand theologically as God's creation. But he soon discovered that natural science "has become a supplier of economic and military power and therefore of political power."[96] Therefore his concern became more and more focused on the practical role natural science plays in our midst, that is, on its political function. In general the power of natural science rests on its veracity. Yet exactly this concept of veracity has always been and still is dubious. Natural science works in an empirical fashion and in so doing makes experience possible by turning to the facts, meaning that which has occurred.[97] What is past is open for the future, because one can connect through laws the possible with what has become fact. This means one can formulate the conditions of possibilities through experience. Yet developing possibilities from facts has become increasingly ambivalent. For instance, in 1939 the German chemist and Nobel laureate Otto Hahn (1879-1968) discovered atomic fission simply by searching for new knowledge. Through his discoveries and without his voluntary contribution, first the military atomic bomb was developed and then, in the long haul, the atomic reactor, which can be viewed as something positive, economically speaking.[98] This example shows that natural science has lost the innocence of its original search for knowledge and must increasingly consider ethical and political issues. Von Weizsäcker himself has done this in an exemplary way. While at the Max Planck Institute he was not just concerned about issues of the environment, but later on more and more with issues of world peace and the coexistence of humanity.

---

96. Von Weizsäcker, *Der Garten des Menschlichen*, 101.
97. Cf. von Weizsäcker, *Der Garten des Menschlichen*, 99.
98. For more detail, cf. von Weizsäcker, *Der Garten des Menschlichen*, 102f.

Paul Davies (b. 1946) is a representative of the present generation of scientists who show a keen interest in elucidating the implications of scientific discoveries for religion in general and theology in particular. After receiving an education in Great Britain and having pursued a teaching career there, he became professor of natural philosophy in the department of physics and mathematical physics at the University of Adelaide in South Australia. He has done extensive research in particle physics and cosmology and has a special gift in interpreting these findings and their implications for religion to the general audience. Though he claims that ten years of radio astronomy have taught humanity more about the creation and organization of the universe than thousands of years of religion and philosophy, he does not want to relegate God to a first cause who got the universe started, since he feels that this is too anthropomorphic a view of the deity.[99]

Modern quantum theory often seems to do away with the notion that everything needs to have a cause and that this first cause is then identified with God.[100] But Davies is uncomfortable with this view. God or some prime cause is still needed to institute the laws that govern the universe. This is an idea that Davies pursues more vigorously in a later publication, *The Mind of God*. God is not the one who initially pushed the button to get everything rolling, but he is the "universal mind existing as part of that unique physical universe."[101] This physical universe is the medium through which God's spirit is expressed. Instead of a supernatural God, Davies claims to opt for a natural one, one that seems to have its affinity in the God of process thought.[102] At the same time Davies is cautious about granting too much insight to the natural sciences. By investigating the workings of nature, a scientist cannot discern anything about God's plan for the world or about the battle between good and evil. Science is reductionistic, and this is its main contribution

99. Cf. Paul C. W. Davies, *Space and Time in the Modern Universe* (Cambridge: University Press, 1977), 217.

100. Cf. Paul Davies, *God and the New Physics* (New York: Simon & Schuster, 1983), 42f. and 216f.

101. Davies, *New Physics*, 223.

102. Cf. Paul Davies, *The Mind of God: The Scientific Basis for a Rational World* (New York: Simon & Schuster, Touchstone Book, 1992), 183, where he admits his affinity to Whitehead, who "replaces the monarchical image of God as omnipotent creator and ruler to that of a participator in the creative process."

to our knowledge of the world. Physics, for instance, does not deal with the questions of the directedness of the creative process or with morals.[103] Therefore religion, according to Davies, is not on the way out.

Whether we are scientists or not, we will still search for a deeper sense of life. "Many ordinary people too, searching for a deeper meaning behind their lives, find their beliefs about the world very much in tune with the new physics."[104] Our worldview has changed so dramatically that the biblical worldview seems out of touch with the way we perceive the world today. This is one of the reasons why theologians should listen to science, since it may provide them with the raw materials for the reconstruction of religious views. Davies thinks this reconstruction is quite advantageous, because he is "fully committed to the scientific method of investigating the world" since "science leads us in the direction of reliable knowledge."[105] While he does not subscribe to "conventional religion," Davies has come to believe more and more strongly, inspired by his scientific work, "that the physical universe is put together with an ingenuity so astonishing" that he "cannot accept it merely as a brute fact" (16). Moreover, he is convinced that sooner or later we all have to accept something as given, whether it is God, logic, a set of laws, or some other premise for our existence. The ultimate questions will always lie beyond the scope of empirical science, since it is there that science and logic will fail us.

Contrary to Einstein, Davies asserts that "God plays dice with the universe" (191). The statistical character of atomic events and the instability of many physical systems with minute fluctuations make it possible for the future to remain open and undetermined by the present. This allows new forms and systems to emerge so that the universe is endowed with freedom for genuine novelty. At the same time, however, Davies shows great admiration for the ontological and cosmological arguments for the existence of God. He is aware that the design argument has been resurrected in recent years by a number of scientists, and he points to the long list of "lucky accidents" and "coincidences" which have allowed life as we know it to evolve (cf. 199). In line with the anthropic principle,

103. So Davies, *New Physics*, 229.

104. Davies, *New Physics*, vii.

105. Davies, *The Mind of God*, 14. The parenthetical numbers in the following text refer to pages in *The Mind of God*.

he toys with the idea that "the apparent 'fine-tuning' of the laws of nature necessary if conscious life is to evolve in the universe then carries the clear implication that God has designed the universe so as to permit such life and consciousness to emerge. It would mean that our own existence in the universe formed a central part of God's plan" (213).

At the same time, he knows that a decision for such a "designer universe" cannot be based on strictly scientific judgment but "is largely a matter of taste" (220). So he concludes with the confession: "I cannot believe that our existence in this universe is a mere quirk of fate, an accident of history, an incidental blip in the great cosmic drama. . . . We are truly meant to be here" (232). Since in the preface of *The Mind of God* he stated that this book is "more of a personal quest for understanding," such a confession at the conclusion is not inappropriate.

The philosopher Ludwig Wittgenstein (1889-1951) confessed at the end of his *Tractatus Logico-Philosophicus:* "We feel that even when *all possible* scientific questions have been answered, the problems of life remain completely untouched." Wittgenstein went on to refer to the mystical.[106] In like manner Paul Davies realizes that there must be something beyond science. He feels that scientific evidence may be interpreted to point beyond science, though this is not a necessary step. With this kind of hesitancy, Davies pushes scientific scrutiny and investigation to its outermost limits while at the same time distinguishing himself from scientists such as Stephen Hawking and Frank Tipler who show no qualms about going beyond that which seems scientifically warranted.

Already in 1978 *Time* magazine named **Stephen Hawking** (b. 1942), professor of theoretical physics at Cambridge University, "one of the premier scientific theorists of the century, perhaps an equal of Einstein."[107] He focuses exactly on those issues which, according to the traditional understanding of the natural sciences, are left out in scientific discourse. As he freely admits, he was motivated in his research in cosmology and quantum theory by questions such as these: "Where did the universe come from? How and why did it begin? Will it have an end, and

---

106. Ludwig Wittgenstein, *Tractatus Logico-Philosophicus,* trans. D. F. Pears and B. F. McGuinness, introduction by Bertrand Russell (London: Routledge & Kegan Paul, 1961), 149 (6.52).

107. According to John Boslough, *Stephen Hawking's Universe* (New York: William Morrow, 1985), 59, in his readable and informative biography.

if so, how?"[108] He rightly claims that these issues are of interest to all of us. To circulate the results of his research to a wider audience, he wrote an informative and popular small book with the title *A Brief History of Time*. He concedes that in 1970 he still thought "that there must have been a big bang singularity" (50). This thesis had won general acceptance. But now he feels that it is perhaps ironic that he has changed his mind and attempts to persuade other physicists to accept exactly the opposite, namely, that in the beginning of the universe there was no such singularity.

Hawking starts with the conviction that natural science has uncovered those laws that, within the limits of the Heisenberg uncertainty principle, tell us how the universe will develop within time, provided we know its state at any one time. According to Hawking, these laws may have initially been decreed by God, but then he seems to have left the universe and no longer intervenes. Yet why has God chosen this original state or configuration of the universe?

> What were the "boundary conditions" at the beginning of time? One possible answer is to say that God chose the initial configuration of the universe for reasons that we cannot hope to understand. This would certainly have been within the power of an omnipotent being, but if he had started it off in such an incomprehensible way, why did he choose to let it evolve according to laws that we could understand? The whole history of science has been the gradual realization that events do not happen in an arbitrary manner, but that they reflect a certain underlying order, which may or may not be divinely inspired. (122)

Hawking wrestles here with the problem that we understand how the universe developed fairly well, but we do not know how this development came about. He surmises that it is very difficult to suppose that from an initial chaotic configuration a universe emerged which in large areas was as uniform as we know it today. To resolve the apparent contradiction between an initial chaos and the subsequent order, Hawking introduces an imaginary time. Imaginary numbers (i) are numbers which when multiplied with themselves still yield negative values, for

108. Stephen W. Hawking, *A Brief History of Time: From Big Bang to Black Holes*, introduction by Carl Sagan (London: Bantam Books, 1988), vi. The parenthetical numbers in the following text refer to pages in this work.

instance, $2i \times 2i = -4$. By connecting our Euclidean understanding of space and time with the quantum theory of gravity, he arrives at the possibility

> in which there would be no boundary to space-time and so there would be no need to specify the behavior at the boundary. There would be no singularities at which the laws of science broke down and no edge of space-time at which one would have to appeal to God or to some new law to set the boundary conditions for space-time. One could simply say: "The boundary condition of the universe is that it has no boundary." The universe would be completely self-contained and not affected by anything outside itself. It would neither be created nor destroyed. It would just BE. (136)

Seen from our real time, we could still assert that our universe was at its smallest 10 or 20 billion years ago, which would correspond to the maximum radius of the history in imaginary time (138). Later, in analogy to a chaotic inflation model, our universe would have expanded. After this phase of expansion it would again slowly contract. Our view of the world would not be that different from the way it used to be. But in relation to imaginary time there were no singularities, neither at the beginning nor at the end.

Hawking knows that the idea of space and time as a surface closed in itself without boundaries has far-reaching consequences for our understanding of how God is related to the universe (for the following, including the quote, see 140f.). Up to now one could still think that God had given the universe certain laws according to which it develops and that through divine intervention he could break through these laws. Since these laws do not tell us what the universe looked like at the beginning, "it would still be up to God to wind up the clockwork and to choose how to start it off. So long as the universe had a beginning, we could suppose it had a creator. But if the universe is really completely self-contained, having no boundary or edge, it would have neither beginning nor end: It would simply be. What place, then, for a creator?"

Hawking says in conclusion that so far scientists have been too busy developing new theories to describe the "what" of the universe that they had no time to ask the why question (for the following, including the quote, see 174f.). On the other hand, philosophers who traditionally have posed the why question had no time to keep pace with the progress in the natural sciences. Therefore, according to Hawking, they have confined themselves more and more to an analysis of language. But once a complete and comprehensive theory is developed, philosophers, scientists, and all other people must participate in the discussion of the question "of why it is that we and the universe exist. If we find the answer to that, it would be the ultimate triumph of human reason — for then we would know the mind of God."

In the tradition of the British apologist William Paley (1743-1805), Hawking compares the function of God with that of a watchmaker. And in analogy to the classical Thomistic understanding of miracles, he understands the present activity of God as an intervention, an interruption or a breakthrough of the ruling laws of nature. This shows in a paradigmatic way that Hawking, similar to many other scientists, has an understanding of God that corresponds neither to biblical understanding nor to current theological formulations. Since Hawking, however, is a natural scientist, one cannot fault him for that. Much more important is another point, namely, that he realizes that scientists can no longer ignore the why question. A simple description of what is or how it happens does not suffice. Especially from an ecological perspective, it has become increasingly urgent to include consequences in our considerations. If we want to find direction in today's world and gain meaning for our existence, we can neither be satisfied with a description of the state of the world and the cosmos nor its trends.

In line with the British skeptic tradition, Hawking resolved the why question in his own manner. There is no mysterious initial singularity, but the states prior to and after the "beginning" merge into each other. The universe is closed in itself. If we were to adopt this view, we would return to the closed causal-mechanistic worldview of the nineteenth century. Yet at the end Hawking admits that the why question is not resolved. He still hopes for a complete theory as a future possibility. To stay with Hawking's terminology, we still cannot understand the mind of God. For Hawking this is not a mystery which cannot be solved in principle, but rather one that perhaps we can solve.[109] Yet he cannot adduce any justified reasons why such a hope would find its fulfillment. We are confronted here with the basic question of why something is and

109. Cf. Barbour, *Religion,* 140.

not simply nothing. Hawking has pushed back the vexing questions concerning creation one step further, but he has not solved them. Theologically speaking, this should not surprise us. Hawking is motivated by why questions. In pursuing them he necessarily comes up against the limits of that which science can assert.

Frank J. Tipler (b. 1947), professor of mathematical physics at Tulane University in New Orleans, wrote *The Anthropic Cosmological Principle* with John D. Barrow (b. 1952). In this book he wanted to introduce a new teleology on a scientific basis. Then he wrote the voluminous book *The Physics of Immortality,* in which he picks up on classical American Deism and transforms it to a rational Christian faith. In unfolding the anthropic principle, Tipler first rejects the fundamental presupposition that humanity as an observer in the universe should not have a privileged position. Instead he shows that an observer can only occupy that place at which the conditions existed for his evolution and existence so that he could emerge. Such a place must necessarily be something special. Tipler refers to the thesis of the British theoretical physicist Brandon Carter (b. 1942), first introduced in 1974, which at that time Carter already called the anthropic principle. Carter claimed "that what we can expect to observe must be restricted by conditions necessary for our presence as observers. (Although our situation is not necessarily *central,* it is inevitably privileged to some extent.)"[110]

Tipler and Barrow start with the statement that there are a number of very unlikely — and coincidental — accidents which are totally independent from each other. These accidents seem to be necessary if an observer, whose basic building material is carbon compounds, is to emerge in our universe.[111] This leads them to establish three different anthropic principles. First there is the weak anthropic principle which states: *"The observed value of all physical and cosmological quantities are not equally probable, but they take on values restricted by the requirement that there exist sites where carbon-based life can evolve and by the requirement that the Universe be old enough for it to have already done so"* (16). The strong anthropic principle then goes

110. Brandon Carter, "Large Number Coincidences and the Anthropic Principle in Cosmology," in *Confrontation of Cosmological Theories with Observational Data,* ed. M. S. Longair (Dordrecht: D. Reidel, 1974), 291.

111. Cf. John D. Barrow and Frank J. Tipler, *The Anthropic Cosmological Principle,* 2nd ed. (Oxford: Clarendon Press, 1988), 5, where they follow Carter's lead. The parenthetical numbers in the following text refer to pages in Barrow and Tipler's work.

one step further and claims: *"The Universe must have those properties which allow life to develop within it at some stage in its history"* (21). Lastly, they formulate the final anthropic principle, which Tipler has expanded in his second book and which states: *"Intelligent information-processing must come into existence in the Universe, and, once it comes into existence, it will never die out"* (23). How far Tipler and Barrow go beyond the strict limits of physics at this point, becomes visible when they claim on the basis of the strong anthropic principle that it is incomprehensible why life should become extinct once it has evolved but has not yet decisively influenced the universe at large. Tipler and Barrow emphasize the necessity of a human observer. In classical physics, humanity had an ancillary role in the universe. But in modern physics, especially in the Copenhagen interpretation of quantum mechanics, the human observer plays a very decisive role (cf. 458).

Two other things are important for Tipler and Barrow: First, the possibility of space travel. Humanity is no longer a fringe phenomenon in the immense depth of the universe. At least in principle humanity can gradually colonize essential parts, if not all, of the visible cosmos (cf. 613f.). The second thing which has impressed Tipler and Barrow is computer technology. In analogy to that technology, humanity does not consist of body and soul but "a program designed to run on a particular hardware called a human body" (659). Such a program can then be duplicated and transferred at will. Having stated this, the authors conclude the anthropic principle with a similar scenario, as Tipler's more recent book, *The Physics of Immortality*, shows. The final point of development, meaning the omega point, is attained once "life will have gained control of *all* matter and forces not only in a single universe, but in all universes whose existence is logically possible; life will have spread into *all* spatial regions in all universes which could logically exist, and will have stored an infinite amount of information, including *all* bits of knowledge which is logically possible to know. And this is the end" (677).

While the weak anthropic principle only states the obvious, namely, that only under certain conditions can life exist and that without these conditions it could not have existed, the strong and the final anthropic principles start with presuppositions that are not without problems.[112]

---

112. Cf. also the objections made by Willem B. Drees, *Beyond the Big Bang: Quantum Cosmologies and God* (La Salle, Ill.: Open Court, 1990), 78f.

The strong anthropic principle presupposes that certain initial data must exist, while the final anthropic principle is a mere postulate on the basis of the strong anthropic principle. In his book simultaneously published in the USA and Germany, *The Physics of Immortality*, Tipler has given up any theological restraints and immediately claims in the preface that "physicists can infer by calculation the existence of God and the likelihood of the resurrection of the dead to eternal life in exactly the same way as physicists calculate the properties of the electron."[113] He sees his venture into theological terrain justified, since in the twentieth century theologians have largely withdrawn from questions about nature and the cosmos. Wolfhart Pannenberg is "a very rare exception" because he is "one of the very few modern theologians to truly believe that physics must be intertwined with theology" (xxiii-xxiv). But Tipler immediately reveals his reductionistic inclinations when he emphasizes: "It is necessary to regard all forms of life — including human beings — as subject to the same laws of physics as electrons and atoms. I therefore regard a human being as nothing but a particular type of machine, the human brain as nothing but an information processing device, the human soul as nothing but a program being run on a computer called the brain" (xi).

In *The Physics of Immortality*, Tipler wants to describe his omega point theory as "a testable physical theory for an omnipresent, omniscient, omnipotent God who will one day in the far future resurrect every single one of us to life forever in an abode which is in all essentials the Judeo-Christian Heaven" (1). Such proof is indeed an immense program for a physicist. Tipler admits, however, that his omega point theory is "a viable scientific theory of the future of the physical universe, but the only evidence in its favor at the moment is the theoretical beauty, for there is as yet no confirming experimental evidence for it" (305). Tipler believes that the odds are quite high that the omega point theory is true. He calls himself neither a Deist, meaning, according to him, that he does not believe in a personal creator God who affects the world from outside it, nor a Christian, since he neither accepts the resurrection of the dead nor the Trinity nor the real presence of Christ in

---

113. So Frank J. Tipler, *The Physics of Immortality: Modern Cosmology, God, and the Resurrection of the Dead* (New York: Doubleday, 1994), ix. The parenthetical numbers in the following text refer to pages in this work.

the Lord's Supper. More appealing to him is American Deism, though he admits that its God is too impersonal. According to Tipler, "religion can be based on physics only if the physics shows that God *has* to be personal, and further, that the afterlife is an absolutely solid consequence of the physics" (327).

Tipler's main goal is to bridge the ugly, broad ditch between physics and religion. He considers it a grave problem that, for many theologians and natural scientists, religion and science have virtually nothing to do with each other, that religion concerns itself primarily with moral issues and science with facts. But according to Tipler, moral decisions, too, must be secured by facts. "If religion is permanently separated from science, then it is permanently separated from humanity and all of humanity's concerns. Thus separated, it will disappear" (332). Tipler therefore opts for an integration of theology into physics. He treats theological assertions as purely scientific assertions so that theology becomes for him "a branch of astronomy," and like any other science it is based only on reason and no longer on revelation (336). "Theology is nothing but physical cosmology based on the assumption that life as a whole is immortal" (for this and following quote, see 338f.). Science can offer the same consolation in the face of death that religion once offered. "Religion is now part of science."

Unlike those scientists so far reviewed, Frank Tipler has not only touched the limits that the natural sciences impose upon themselves but he has gone far beyond these limits. Indeed, it is questionable whether one must distinguish between knowledge about the world and knowledge about salvation, if God is ultimately the creator of the world and mediator of salvation. Wolfhart Pannenberg, too, refused to distinguish between salvation history and world history. If Einstein, however, is right, that the parameters for our orientation in the world — space, time, and matter — have no absolute validity but are only defined in relation to each other, then we wonder how we can discern within this world one or several parameters that guide us in our conduct and show us the way we should take. The evolving universal spirit which, according to Tipler, emerges at the end of the universe is nothing other than what was potentially already present at the beginning. Therefore it only points out a factual direction, but not a preferential one. The transition from the weak anthropic principle which actually is a tautology to the strong anthropic principle presupposes fundamental decisions which are not justified by the mere existence of the universe itself. Tipler is not just a detached, objective, and descriptive scientist, but he is also an implicit theologian. He determines that which is desirable and what makes sense to him. At this point the discussion begins with theology, because natural science alone cannot ultimately sanction such valuations.

e. Theological Positions Today (Barth and Torrance, Teilhard and Process Thought, Pannenberg and Moltmann)

When we now attempt to delineate the positions assumed by theology today, we must first start with the classical position of **Karl Barth** (1886-1968). Like no one else, he decisively influenced the theological climate in the first half of the twentieth century. As one of the founders and representatives of the so-called neo-Reformation theology, Barth a priori precluded in his doctrine of creation any contact or dialogue with the natural sciences. In epical breadth he unfolded this doctrine in more than 2,200 pages. It is divided in four chapters, with each chapter constituting a separate volume: *The Work of Creation; The Creature; The Creator and His Creature;* and finally *The Command of God the Creator.* While the title of the last volume shows ethical intentions, at least in the other three one would assume some discussion with the natural sciences.

Already in the first volume, however, we notice a strictly theological bent. Creation is understood in view of the covenant. In the preface Barth preempts any potential criticism of his procedure when he writes:

> It will perhaps be asked in criticism why I have not tackled the obvious scientific question posed in this context. It was my original belief that this would be necessary, but I later saw that there can be no scientific problems, objections or aids in relation to what Holy Scripture and the Christian Church understand by the divine work of creation.... There is free scope for natural science beyond what theology describes as the work of the Creator. And theology can and must move freely where science which really is science, and not secretly a pagan *Gnosis* or religion, has its appointed limit.[114]

114. Karl Barth, *Church Dogmatics,* vol. 3, *The Doctrine of Creation,* pt. I, ed. G. W. Bromiley and T. F. Torrance (Edinburgh: T. & T. Clark, 1958), ix-x.

If we fear that with this dictum any discussion with the natural sciences is ultimately precluded, we may regain hope with the next sentence: "I am of the opinion, however, that future workers in the field of the Christian doctrine of creation will find many problems worth pondering in defining point and manner of this twofold boundary."

A theologian is only concerned with God and God's work, while a scientist researches the world. But both refer to the created order and interpret it either as world or as creation. Therefore one would assume that a debate should ensue between theology and science. Yet Barth wants nothing to do with a dialogue because, as he states later in the same volume, we are confronted with "a fundamental difference between the Christian doctrine of creation and every existent or conceivable world-view."[115] The Christian doctrine of creation cannot become a worldview, nor can it be supported by one nor support one. "It cannot come to terms with these views, adopting an attitude of partial agreement or partial rejection. . . . Its own consideration of these views is carried out in such a way that it presents its own recognition of its own object with its own basis and consistency, not claiming a better but a different type of knowledge which does not exclude the former but is developed in juxtaposition and antithesis to it." Barth once more makes it clear that in the strict sense no dialogue is possible with other worldviews, including the natural sciences.

Theology and natural science work and argue alongside each other; they are phenomena which stand strictly parallel to each other. Barth enters this way not out of ignorance or laziness. In the second part of his *Doctrine of Creation* he extensively deals with theological apologetics, especially with Otto Zöckler, and shows that it attempted to defend the special place of humanity in the world over against Darwin's theory of descent. Barth has some good things to say about this attempt, but he contends that we should not deceive ourselves, "that we have attained to real man, to his uniqueness in creation."[116] One is left with the phenomenon of humanity, but it does not reach its reality as shown to us by God in Jesus.[117] To talk about creation means for Barth to talk about God and his

115. Barth, *Church Dogmatics*, vol. 3, pt. I, p. 343, for this quotation and the following.

116. Karl Barth, *Church Dogmatics*, vol. 3, *The Doctrine of Creation*, pt. II, ed. G. W. Bromiley and T. F. Torrance (Edinburgh: T. & T. Clark, 1960), 94.

117. Barth begins each section of his anthropology with some reflections on Jesus; i.e., he begins each section with a christological foundation.

activity, and to talk about humanity means to talk about Jesus Christ, since in him we encounter the creature chosen and elected by God.

Barth's strictly theological understanding of creation has certainly been helpful and liberating for many, because he has shown theologians and laypeople a space in which they can unfold the theological doctrine of creation without being disturbed by scientific knowledge. Certainly this also helped scientists, who saw their faith in creation threatened by scientific knowledge. At the same time, however, the doctrine of creation lost its anchorage in the world, a world which is largely shaped by applied science (i.e., by technology). This led to a standstill in the dialogue, and in a society shaped by the natural sciences, theology was increasingly considered to be on the fringe. Yet some followers of Barth nevertheless vigorously engaged in a dialogue with the sciences. Representing scientists is Günter Howe, whom we have already mentioned. On the side of the theologians we must note especially Thomas Torrance.

Thomas F. Torrance (b. 1913) was for many years professor of Christian dogmatics at the University of Edinburgh. Though in many ways decidedly influenced by Barth, he has a keen interest in science, if for no other reason than that for him theology is scientific too.

> There are not two ways of knowing, a scientific way and a theological way. Neither science nor theology is an esoteric way of knowledge. Indeed because there is only one basic way of knowing we cannot contrast science and theology, but only natural science and theological science, or social science and theological science. In each we have to do with a fundamental act of knowing, not essentially different from real knowing in any field of human experience. Science is the rigorous and disciplined extension of that basic way of knowing and as such applies to every area of human life and thought.[118]

"Theology is the unique science devoted to the knowledge of God, differing from other sciences by the uniqueness of its object which can be apprehended only on its own terms and from within the actual situation it has created in our existence in making itself known."[119] Like all

---

118. Thomas F. Torrance, *God and Rationality* (New York: Oxford University Press, 1971), 91.

119. Thomas F. Torrance, *Theological Science* (New York: Oxford University Press, 1969), 281.

other sciences, theological science is a human inquiry. It is not quackery, but it will put all claims that purportedly are the results of theological thinking to the most severe test to make sure it does not miss its object matter, the knowledge of God.

Since the primary object of theological inquiry is the one God who is the source of all being and the ground of all truth, theology is concerned with wholeness and unity, which may set it apart from any other science.[120] Like other scientific disciplines, theological science refers to the externally given reality. Moreover, theology comes up in its investigations against a boundary beyond which it cannot penetrate and which it cannot pass without inconsistency and error. Even theological science has its limits. Yet in theology we have to do with a divine object that draws us to itself. From the very beginning our knowledge of God starts with a union and not a disjunction between subject and object. This does not mean that subjectivism reigns supreme in theology, but that we could never talk about God unless God first speaks to us. This also means for Torrance that a strictly natural theology is not possible.

Contrary to Barth, Torrance is not afraid of using the term "natural theology." Yet he considers natural theology to be incomplete in itself and only consistent if it is coordinated with positive theology, meaning that it has its proper place within "the embrace of the theology of God's self-revealing interaction with us in the world."[121] Its concepts and theorems lack meaning and cogency in themselves and only make sense when supplemented and interpreted from the level of divine revelation. The reason for his acceptance of natural theology lies in the fact that "theology by its very nature can be pursued only within the rational structures of space and time within which we are placed by God, through which he mediates to us knowledge of himself, and within which we may develop and articulate our knowledge of him."[122] This means that theology is an enterprise within this world and must do its own business with continuing reference to the parameters which we also know from the other sciences.

It is no surprise that a doctrine of creation is of utmost importance

120. Cf. Torrance, *Theological Science*, 282.

121. Thomas F. Torrance, *Reality and Scientific Theology* (Edinburgh: Scottish Academic Press, 1985), 64.

122. Torrance, *Reality and Scientific Theology*, 64.

for Torrance. Two items are significant here: God in his transcendent freedom made the universe out of nothing; and in giving it a reality distinct from his own but dependent on him, he endowed the universe with an immanent rationality making it determinate and knowable. Theologically this means that the world can be understood as existing because of God's creative work and that it is being maintained by God. Considered from the point of view of natural science, however, the world can be known and understood because of its immanent rationality and determinate character. Nature then can only speak ambiguously of God if taken by itself. Though "it may be interpreted as pointing intelligibly beyond itself to God, it does not permit of any necessary inferences from its contingence to God."[123] There is no inference possible from the created to the creator. It is not possible to prove in a rational way that God exists and that he has created the world, since the rational connection between creation and God is grounded in God alone. A natural theology or a theology of creation that does not take its starting point with God and his self-disclosure is doomed to failure.

The Christian doctrine of God who is the creator of an orderly universe, who brought it into existence out of nothing and continuously preserves it from falling back into chaos and nothingness, has significance far beyond the confines of theology. All empirical-theoretical inquiry rests upon this contingent character of the world. "Natural science assumes the contingence as well as the orderliness of the universe."[124] While the notion of contingence was already present in Greek thought, it was restricted there by the notion of the created as being the embodiment of divine reason as well as by the dualism between eternal form and accidental matter. This meant that for ancient natural science the empirical was regarded as something secondary. For this reason it took so long for the contingent and empirical character to gain the upper hand in the development of science. This means that natural science with its modern emphasis on the factual and empirical is largely due to the Judeo-Christian understanding of creation.

Since modern natural sciences and the Judeo-Christian tradition

---

123. Thomas F. Torrance, *Space, Time, and Incarnation* (London: Oxford University Press, 1969), 59f.

124. Thomas F. Torrance, "God and the Contingent World," *Zygon* 14 (December 1979): 329.

are historically interdependent, Torrance can assert: "But since it is the contingence of the realities of the empirical universe upon God that gives them their intelligibility and enables us to grasp their natural and inherent structures, genuine interaction between theological science and natural science cannot but be helpful to both."[125] God is the creator of all things visible and invisible and the source of all rational order in the universe. Therefore both theological science and rational science operate within the same space-time continuum, and theological interpretation and explanation cannot properly take place without constant dialogue with natural science.[126] There are especially two points at which dialogue becomes necessary for theology: in its emphasis on incarnation and resurrection as its basic and all-embracing miracles upon which the Christian gospel rests, because at these points God had acted decisively within the natural order of things.

It becomes clear that Torrance does not want to relegate theology and science into two entirely separate realms. Natural science is concerned with the universe in its natural, contingent process, and theological science focuses on the acts of God which in creation brought those processes into being out of nothing and established them in their utter contingency. Both have to do with the created order, theology inquiring into its transcendent source and ground, and science researching the contingent nature and pattern of that order.[127] There are especially three points which theology should take note of in its conversations with science. (1) There has occurred a basic change in the concept of reality. Here Torrance points quite often to the Copenhagen interpretation of quantum theory, saying: "The emergence of relativity theory has had to give way to a profounder and more differential view of reality in which energy and matter, intelligible structure and material content, exist in mutual interaction and interdetermination."[128] We must take note of (2) the relational concept of space and time and of (3) the multileveled structure of human knowledge. The various sciences can be regarded as constituting a hierarchical structure of levels of inquiry which are open to wider and more comprehensive systems of knowl-

125. Torrance, "God," 346.
126. Cf. Thomas F. Torrance, *Space, Time, and Resurrection* (Edinburgh: Handsel Press, 1976), 22f., for the following.
127. Cf. Torrance, *Space, Time, and Resurrection*, 180.
128. Torrance, *Space, Time, and Resurrection*, 185.

edge. For Torrance this opens up the possibility of a fruitful dialogue without asserting an either-or mentality which excludes the assertions of either theology or science. Torrance is open to accepting the most recent insights of the natural sciences and can also engage in a fruitful dialogue with philosophers of science, such as Michael Polanyi (1891-1976) and Alfred North Whitehead (1861-1947).

While Barth and those who followed him talked about creation strictly from a theological perspective, never venturing to the other side of the fence, there are representatives of another school of thought who freely cross the border and pursue the dialogue on the enemy's own terrain, so to speak. Often they are thereby accused of losing themselves in the scientific quagmire, and their theological positions become questionable. Ranking most prominently among them is the French Jesuit and paleontologist **Pierre Teilhard de Chardin** (1881-1955), whose work became known on the Protestant side in the seventies. The official Roman Catholic Church contributed to his late and posthumous fame. During most of his life it had cast doubts on the theological orthodoxy of his views, and only at the Second Vatican Council (1962-65) was he finally rehabilitated.

Teilhard is both a scientist and a theologian and priest. In his self-understanding the latter gains the upper hand, because he endeavors to find a new synthesis of his theological, physical, paleontological, and paleoanthropological findings with regard to a Christian view of the universe and of humanity.[129] His theological starting point is the incarnation. God has entered this world and will finally unite the world with himself. The universal Christ found by Teilhard in the New Testament is "the organic center of the entire universe."[130] He is the Alpha and Omega, the beginning and end. If Christ is universal, Teilhard concludes, then redemption and the fall "must extend to the whole universe" and assume cosmic dimensions.[131]

Teilhard wants to overcome the old static dualism between spirit and matter. He thinks this can be accomplished by viewing spirit and

129. So very clearly Ernst Benz, *Evolution and Christian Hope: Man's Concept of the Future, from the Early Fathers to Teilhard de Chardin,* trans. H. G. Frank (Garden City, N.Y.: Doubleday, Anchor Book, 1968), 212.

130. Pierre Teilhard de Chardin, "Note on the Universal Christ," in *Science and Christ,* trans. René Hague (New York: Harper & Row, 1965), 14.

131. Teilhard, "Note," 16.

matter together in a universe which is historically advanced through an inward-guided evolution. There are four stages of development: the cosmosphere as the origin of the cosmos; the biosphere as the advancement of life; the noosphere as "the Earth's thinking envelope,"[132] which is intensified through a psychophysical convergence, an event Teilhard calls the planetization; finally, the Christosphere emerges when the whole cosmos is permeated by Christ and taken up in God. Important for this is the omega point. It is, so to speak, the inner circle of the universe which describes the end point of the development. God incarnate is reflected in our noosphere. He is the reflection "of the ultimate nucleus of totalization and consolidation that is biopsychologically demanded by the evolution of a *reflective* living Mass."[133] As the divine converges with the material, the universe becomes personalized and the person (of Christ) becomes universalized by merging with all of humanity and all that is material.[134] The whole evolutionary activity is therewith centered in a process of union or communion with God.

For Teilhard religion and natural science are not opposites. "Even though the various stages of our interior life cannot be expressed strictly in terms of one another," they must nevertheless "agree in scale, in nature and tonality."[135] Science "uses certain exact 'parameters' to define for us the nature and requirements, in other words the physical stuff, of 'participated' being. It is these parameters that must in the future be respected by every concept of Creation, Incarnation, Redemption and Salvation." Teilhard does not want a gnostic symbiosis between theology and science, but he is convinced that a convergence is coming, because theology and science internally hang together. He regards the scientific study of the world as an analytic pursuit that follows a direction which leads away from the divine reality.[136] At the same time, this scientific insight also shows a synthetic structure of the world and forces us to change our direction so that we turn back to the unique center of all

132. Cf. Pierre Teilhard de Chardin, *The Heart of Matter*, trans. René Hague (New York: Harcourt Brace Jovanovich, 1978), 31.

133. Teilhard, *The Heart of Matter*, 39.

134. Cf. Teilhard, *The Heart of Matter*, 44f.

135. For this and the following quote, cf. Teilhard, *Science and Christ*, 221ff., in a letter of November 2, 1947.

136. So Pierre Teilhard de Chardin, "Science and Christ or Analysis and Synthesis," in *Science and Christ*, 21.

things which is the Lord our God. Christians need not be afraid of the results of scientific research, whether in physics, biology, or history.[137] Often the analyses in the natural and historical sciences are correct. Yet they do not threaten the Christian faith. "Providence, the soul, divine life, are synthetic realities. Since their function is to 'unify,' they presuppose, outside and below them, a system of elements; but those elements do not constitute them; on the contrary, it is to those higher realities that the elements look for their 'animation.'" Science does not endanger faith, but helps us to know God better, to better understand and appreciate God. It does not make sense and it is unfair to pit science and Christ against each other or to separate the two into different realms which are foreign to each other.

On their own the natural sciences cannot discover Christ. The scientific endeavor gives birth to a yearning which Christ satisfies. Teilhard celebrates a cosmic Eucharist in a visionary or even mystical-ecstatical way. In this process the divine fire permeates and illuminates the whole cosmos. However, Teilhard's idea of a Christification of the universe leads one to wonder if he has not used the traditional Roman Catholic understanding of transubstantiation as an unreflected premise.[138] It certainly may have influenced him. But much more important for Teilhard is the idea of evolution, which he understands in a totally christocentric way. Similar to Karl Heim, Teilhard starts with the presupposition that there is no opposition between faith and thought, since both are gifts of God. Therefore, in an apologetic way Teilhard wants to issue a reasonable invitation to faith.[139] He does not start with physics, as does Heim, but with biology. It is his special contribution that he reaches out into that area about which the Roman Catholic Church has been suspicious for so long, and he attempts to penetrate it theologically.

Like Teilhard, **Alfred North Whitehead** attempted to combine in his own personal life his interests in the natural sciences with a religious outlook on the world. As a philosopher and mathematician he taught at Cambridge University and at the Imperial College of Science and

---

137. For the following, including the quote, cf. Teilhard, "Science and Christ," 35.

138. Cf. Pierre Teilhard de Chardin, *The Divine Milieu: An Essay on the Interior Life* (New York: Harper Torchbook, 1965), 142ff.

139. Cf. also Henri de Lubac, *The Religion of Teilhard de Chardin,* trans. René Hague (London: Collins, 1967), 232.

Technology at London, before he was invited in 1924 to a professorship in philosophy at Harvard University. He attempted to produce a comprehensive metaphysical system which would also take account of scientific cosmology since he was convinced that religion and natural science are so important for humanity "that the future course of history depends upon . . . the relations between them."[140]

The significance of religion is not accidental for Whitehead. It was only "the medieval insistence on the rationality of God, conceived as with the personal energy of Jehovah and with the rationality of a Greek philosopher," that provided the motivation for research into nature (18). The philosophy of Aristotle gave one a clear head and an analytic mind, while the Judeo-Christian tradition provided the conviction that every detail of nature was supervised. Research into nature could only result in the vindication of faith in rationality. In other religions God was either conceived as too arbitrary or as too personal to generate the possibility of a natural science. What we understand today as modern science did not come about by itself. "The faith in the order of nature which has made possible the growth of science is a particular example of a deeper faith. This faith cannot be justified by any inductive generalization. It springs from direct inspection of the nature of things as disclosed in our own immediate present experience" (27). It is not simply a primordial mystery that Whitehead posits at the depth or an Aristotelian prime mover who would initiate a process. For Whitehead God is the "Principle of Concretion" (250). This means that out of the wealth of certain events or things come concrete possibilities by virtue of their inclusion and exclusion in the process of discrimination. "God is the ultimate limitation, and His existence is the ultimate irrationality. For no reason can be given for just that limitation

---

140. Alfred North Whitehead, *Science and the Modern World: Lowell Lectures, 1925* (New York: Macmillan, 1926), 260; parenthetical page references in the following paragraph are to this work. For a good introduction to the thought of Whitehead, cf. Edward Pols, *Whitehead's Metaphysics: A Critical Examination of "Process and Reality"* (Carbondale: Southern Illinois University Press, 1967). Other excellent introductions to Whitehead are provided by Victor Lowe, Charles Hartshorne, and A. H. Johnson, *Whitehead and the Modern World: Science, Metaphysics, and Civilization. Three Essays on the Thought of Alfred North Whitehead,* preface by A. Cornelius Benjamin (Boston: Beacon Press, 1956), and Paul Arthur Schilpp, ed., *The Philosophy of Alfred North Whitehead,* 2nd ed. (La Salle, Ill.: Open Court, 1991).

which it stands in His nature to impose. God is not concrete, but He is the ground for concrete actuality. No reason can be given for the nature of God, because that nature is the ground of rationality" (257). God is the underlying cause of everything which is not caused by anything else and who provides the possibility that something issues forth from all possible possibilities.

It is important to see the whole world in a continuous process which goes through several layers. First there are "the actualized data presented by the antecedent world," meaning the past, which has already become.[141] Then there are the "non-actualized potentialities" which lie ready to promote their fusion into a new unity of experience. And finally there is "the creative advance" whereby these not-yet-actualized potentialities pass into the future and thereby become present. Whitehead sees this process not as something automatic, but as something in which God is intimately involved.

Whitehead distinguishes between a primordial and a subsequent nature in God. In his primordial nature God is "the unlimited conceptual realization of the absolute wealth of potentiality."[142] This does not mean that God is prior to all creation, but rather that he is with all creation. Apart from God there would be no actual world, since nothing could be actualized; and apart from the actual world with its creativity, "there would be no rational explanation of the ideal vision which constitutes God."[143] Thus God needs the world as his arena of actualization and the world needs God to actualize it. This interdependence becomes even more evident in God's consequent nature. Since all things are interrelated, Whitehead assumes that the world reacts to God. Thus God "shares with every new creation its actual world."[144] While in God's primordial nature all groundwork for the possible world is given, God in his consequent nature provides the weaning of his physical feelings from his primordial concepts through a kind of feedback. Whitehead describes the nature of God's subsequent involvement in

141. Cf. Alfred North Whitehead, *Nature and Life* (Cambridge: University Press, 1934), 60.

142. Alfred North Whitehead, *Process and Reality: An Essay in Cosmology* (New York: Macmillan, 1960), 521.

143. Cf. Alfred North Whitehead, *Religion in the Making: Lowell Lectures, 1926* (New York: Macmillan, 1926), 157.

144. Whitehead, *Process and Reality*, 523.

the world as "the perpetual vision of the road which leads to the deeper realities."[145] Since God's subsequent nature is always moving on and integrates the actualities of the world into the primordial whole which is unlimited conceptual reality, God provides the binding element in the world. He confronts what is actual in the world with what is possible for it, and at the same time provides the means for merging the actual with the possible.

Both God and the world are instruments of novelty for the other. But God and the world move conversely to each other with respect to their processes. As primordially one, God in his consequent nature acquires in the interchange with the world the multiplicity of the actual occasions and absorbs them into his own primordial integrative unity. The world, however, as primordially many, acquires in the interchange with God in his subsequent nature an integrative unity. This integrative unity is a novel occurrence which is absorbed into the multiplicity of its primordial nature. God and world are coaxing each other along. God is completed by the finite and the finite is completed through confrontation with the eternal. Whitehead sums up his ideas by saying: "What is done in the world is transformed into a reality in heaven, and the reality in heaven passes back into the world. By reason of this reciprocal relation, the love in the world passes into the love in heaven, and floods back again into the world. In this sense, God is the great companion — the fellow-sufferer who understands."[146]

Many theologians, philosophers, and scientists have taken up the agenda of Whitehead. A central claim of that agenda is that if religion has any contact with physical facts, "it is to be expected that the point of view of those facts must be continually modified as scientific knowledge advances."[147] We could cite here Charles Hartshorne (1897-2000), who has taught philosophy primarily at the University of Chicago, Emory University, and the University of Texas; or Charles Birch (b. 1918), professor of biology at the University of Sydney, Australia; Schubert Ogden (b. 1928) at the Divinity School of Southern Methodist University; or

145. Whitehead, *Religion in the Making,* 158.

146. Whitehead, *Process and Reality,* 532.

147. Whitehead, *Science,* 271. For a remarkable interdisciplinary attempt of relating Whitehead's philosophy to scientific issues, cf. John B. Cobb, Jr., and David R. Griffin, eds., *Mind in Nature: Essays on the Interface of Science and Philosophy* (Washington, D.C.: University Press of America, 1977).

David Griffin (b. 1939), the executive director of the Center for Process Studies in Claremont, California. Yet the most influential attempt is that by **John B. Cobb, Jr.** (b. 1925), who, as a former student of Hartshorne, has picked up many of Whitehead's insights and applied them to theology. He can truly be called the leading representative of process theology. Of Cobb's many publications, *A Christian Natural Theology Based on the Thought of Alfred North Whitehead* is a deliberate attempt to show how Whitehead's philosophy can be fruitful for advancing the dialogue between theology proper and the sciences. Special attention is given to the possibility of a Christian natural theology. Cobb is convinced that a natural theology is not a dubious luxury in our generation, but serves as a foundation for the proclamation and realization of faith. He uses the work of Whitehead for developing such a Christian natural theology, since it is "a fully developed alternative to the nihilistic tendencies of most modern thought."[148] Cobb is well aware of the pitfalls of traditional natural theology, especially of its portrayal of a God who is impassible and immutable and therefore deeply involved with neither creation nor the affairs of human history. Yet God need not be portrayed as a "transcendental snob."

A Christian theologian must responsibly reflect on the structure and content of faith and therefore needs a trustworthy vehicle for expressing this faith. The theologian has two choices, either to create his own conceptuality or adopt and adapt some already existing philosophy.[149] Cobb freely admits that every natural theology reflects some fundamental perspective on the world, and natural theology is purely the result of neutral objective reason. Since every argument begins with premises and the final premises themselves cannot be proved, it is important that these premises agree with one's own persuasion. "Hence, a Christian theologian should select for his natural theology a philosophy that shares his fundamental premises, his fundamental vision of reality."[150] In this context Cobb opts for the philosophy of Whitehead, since that philosophy has been deeply affected in its starting point by the Christian vision of reality. But Cobb does not unquestioningly follow

---

148. John B. Cobb, Jr., *A Christian Natural Theology Based on the Thought of Alfred North Whitehead* (London: Lutterworth, 1965), 15.

149. Cf. Cobb, *A Christian Natural Theology*, 263.

150. Cobb, *A Christian Natural Theology*, 266.

Whitehead. Though Whitehead advocated a nonspatial understanding of God, Cobb finds it more intelligible to say "that God is everywhere." God is immediately related to every place and there is nowhere we could flee from him.[151]

Like Whitehead, Cobb picks up on the notion of process and rejects the old idea that our world consists of a space-time receptacle within which certain events take place. Cobb argues that energy events themselves are the ultimate reality. These events have patterns of relation with each other which can be described as extensive. Therefore there is a successive and contemporaneous quality to these events. Each event is a subject for itself and an object for its successors. This means there is a vast web of interconnectedness and also a peculiar individualization. This also holds true for humanity.

In obtaining the total picture we must consider God, who influences the process by being what he is. Therefore the future is not only a different connection of previous human experiences. God "seeks to lure the new occasion beyond the mere repetition of past purposes and past feelings or new combinations among them. God is thus at once the source of novelty and the lure to finer and richer actualizations embodying that novelty. Thus God is the One Who Calls us beyond all that we have become to what we might be."[152] God is not totally free in his creative activity, but sees that which is at hand and attempts to bring it in line with the aim he has for us. Though that kind of progress toward a full humanity is rather slow, Cobb assures us that God has brought us a long way.

While Whitehead's philosophical insights may not provide a panacea for every task facing theology, many points in his system nevertheless commend themselves to us.[153] It succeeds in picking up the dynamic worldview of twentieth-century natural science without forfeiting faith in a God who cares. It is also good to hear that God does not steamroll over our own concerns and efforts, but considers each and every individual event. Yet we wonder whether divine persuasion alone is sufficient to bring this world to its hoped-for conclusion. The

151. John B. Cobb, Jr., *God and the World* (Philadelphia: Westminster, 1969), 77.

152. Cobb, *God and the World*, 82.

153. Cobb, *A Christian Natural Theology*, 277ff., seems too optimistic when he suggests that Whitehead offers a solution to virtually every task of theology from confessional loyalty to interreligious dialogue.

immensity of evil and human depravity loom too large to allow for such a divine transformation.[154]

In Wolfhart Pannenberg and Jürgen Moltmann we are dealing with two theologians who have a keen awareness of the natural world and who deal with that world from their own theological vantage point. **Wolfhart Pannenberg** (b. 1928) has dealt only incidentally with topics explicitly pertaining to the natural sciences.[155] But already in his "Dogmatic Theses on the Doctrine of Revelation" (1961) he rejected a special salvation history and relegated theology to the same realm as any other science. He claimed that "historical revelation is open to anyone who has eyes to see."[156] Pannenberg therefore opens up his theological method for everybody to investigate, and he is ready to allow for a falsification through the natural sciences.[157] Pannenberg wants theology to be included — as an equal partner — in the community of rational scientific investigation of the world. In anthropology, for instance, his goal is "to lay theological claim to the human phenomena described in the anthropological disciplines. To this end, the secular description is accepted as simply a provisional version of the objective reality, a version that needs to be expanded and deepened by showing that the anthropological datum itself contains a further and theologically relevant dimension."[158] Pannenberg does not simply isolate certain facts and make them fruitful as a foundation for revelation. Pannenberg goes further. He is convinced that "when the events of nature and history events are properly understood, in and of themselves, knowledge of their being rooted in God and God's will is conveyed."[159] But complete knowledge is only available once history has come to its end.

154. Cf. my caution against this optimism of process theology in Hans Schwarz, *Evil: A Historical and Theological Perspective* (Minneapolis: Fortress, 1995), 184.

155. For an assessment of Pannenberg's involvement with science, cf. the articles by Gregory R. Peterson, John Polkinghorne, and Stanley J. Grenz, "What Shall We Make of Wolfhart Pannenberg?" *Zygon* 34 (March 1999): 139-66.

156. Cf. Wolfhart Pannenberg, "Dogmatic Theses on the Doctrine of Revelation," in *Revelation as History,* ed. Wolfhart Pannenberg, trans. David Granskou (New York: Macmillan, 1968), 135, in thesis 3.

157. So Philip Hefner, "The Role of Science in Pannenberg's Theological Thinking," in *The Theology of Wolfhart Pannenberg,* ed. Carl E. Braaten and Philip Clayton (Minneapolis: Augsburg, 1988), 284.

158. So Wolfhart Pannenberg, *Anthropology in Theological Perspective,* trans. Matthew J. O'Connell (Philadelphia: Westminster, 1985), 19f.

159. Hefner, "Role of Science," 269.

In his comprehensive treatise on the theory of knowledge, Pannenberg endeavors to show the scientific character of theology. He claims that theology is a science of God which can approach its object matter only indirectly through the study of religions. To fulfill its scientific character, theological assertions must meet three criteria. First, they must "have a cognitive character"; this means that they must "say something about a state of affairs for which they claim truth."[160] Second, these assertions must be coherent. They have to refer to one object matter. This object matter is given in the indirect self-communication of the divine reality. This divine self-communication occurs in the preliminary experiences of the totality of the reality of meaning as they are present in the religious traditions of faith. Finally the assertions must be open to examination. Pannenberg introduces a preliminary verification about the truth claim. A final verification cannot be established either positively or negatively within the process of history as long as this process is not yet closed. This means that even the assertions of truth in the natural sciences are preliminary and not essentially different from theological claims of truth. In his 1970 essay "Contingency and Natural Law," Pannenberg emphasizes that a common ground upon which natural science and theology can meet without denying their specific differences is that of contingency and order.[161] The Judeo-Christian understanding of God was always decisively influenced by contingent historical events. "New and unforeseen events take place constantly that are experienced as the work of almighty God." One also recognized an orderliness which was dependent on the contingent activity of God.

Another area in which Pannenberg touches the natural sciences is with the concept of "field." Especially in developing his understanding of the doctrine of God as spirit does he use the concept of field as it is commonly understood in physics and applies it to theology.[162] Pannenberg regrets that traditionally spirit is associated with intellect. In the

160. Wolfhart Pannenberg, *Theology and Philosophy of Science*, trans. Francis McDonagh (London: Darton, Longman & Todd, 1976), 327.

161. Cf. Wolfhart Pannenberg, "Contingency and Natural Law," in Wolfhart Pannenberg, *Toward a Theology of Nature: Essays on Science and Faith*, ed. Ted Peters (Louisville: Westminster/John Knox, 1993), 76, including the following quote.

162. Cf. for the following, Wolfhart Pannenberg, "Theological Appropriation of Scientific Understandings: Response to Hefner, Wicken, Eaves, and Tipler," *Zygon* 24 (1989): 256f.; quote on 258.

Old Testament, however, the term "spirit" more frequently denotes a breeze or wind. By equating the spirit of God with the human spirit, there occurred an excessive anthropomorphism in understanding the divine reality. To counteract this tendency Pannenberg takes over the concept of field. According to Pannenberg, "spirit is rather a kind of force, comparable to the wind, but prior to bodily phenomena. If theology wants to be true to the biblical witness, the concept of God as spirit has to be disentangled from the customary identification with mind, an identification which entails an all-too-facile image of God as 'personal.'"

When Pannenberg specifically refers to creation, he again uses the concepts of contingency and field. The theological assertion that the world is contingent on an act of divine creation implies, according to Pannenberg, the assertion "that the existence of the world as a whole and of all its parts is contingent."[163] The world as a whole need not exist at all, but it owes its existence to the free activity of divine creation. This is also true for each part of the world. The contingency of the world shows a close connection between creation and its preservation. The world was not just once called into existence, but each created creature must be preserved in every moment in its existence if it is not to perish. In the Christian tradition such preservation is nothing but continuous creation. Creation occurred not just at the beginning but is verified anew in each moment.

Pannenberg uses field theory, for instance, to translate assertions about angels into our present-day conceptual world.[164] Traditionally angels are considered to be immaterial spiritual realities and powers which, in distinction to the divine spirit, are limited realities. They either verify themselves as God's messengers or they oppose God in demonic freedom. In a field structure angels could be interpreted as the appearance of relatively independent parts of a cosmic field. If one regards the background of the biblical language which talks about angels as personal realities, then one could very well think of them as fields of power or dynamic spheres. As such they could be experienced either as good or evil. Of course, Pannenberg knows that with these deliberations he deviates from the traditional doctrine of angels, but with these meta-

163. Wolfhart Pannenberg, "The Doctrine of Creation and Modern Science," *Zygon* 23 (1988): 8.

164. For the following, Pannenberg, "Doctrine of Creation," 14f.

phors he attempts to grasp that which is theologically essential when we speak about angels.

Pannenberg is convinced that "the theological doctrine of creation should take the biblical narrative as a model in that it uses the best available knowledge of nature in its own time in order to describe the creative activity of God. This model would not be followed if theology simply adhered to a standard of information about the world which has become obsolete long ago by further progress of experience and methodical knowledge."[165] Pannenberg wants to express faith in creation, with the help of modern scientific knowledge, in such a way that the biblical narrative only provides a model for that which ought to be expressed. The content of our faith is thereby largely filled in by our contemporary knowledge. In so doing he can be critical, for instance, with respect to the Priestly creation account which places the creation of the stars relatively late. At the same time he notes astounding analogies in the first chapter of the Bible between our present ideas of the origin and development of the world with that of the people of antiquity. Important for him is not so much the *how* of creation, something that, considering the relativity of its knowledge, modern science can express best, but the *that* of creation. Therefore he does not discuss the assertions of an alternative creationistic science, but simply states: "Theology has to relate to the science there *is* rather than invent a different form of science for its own use."[166] Pannenberg totally relies on the results of modern science. He is confident that truth cannot be divided and that there is no opposition between theology and the natural sciences.

The Tübingen theologian **Jürgen Moltmann** (b. 1926) confesses that, for a long time, he had wanted to write a theology of creation since he has been considered a neo-Reformation theologian and put into the same corner as the young Karl Barth.[167] From very early on we can discern in his writings an ecologically interested understanding of nature and creation. He claims, for instance, that world history and salvation history cannot be separated from each other, "for it is impossible to hand over to condemnation the history that we create through the

165. Pannenberg, "Doctrine of Creation," 19.

166. Pannenberg, "Doctrine of Creation," 7.

167. Cf. Jürgen Moltmann, *God in Creation: A New Theology of Creation and the Spirit of God,* trans. Margaret Kohl (Minneapolis: Fortress, 1993), in his preface to the paperback edition, xi.

knowledge of nature and technology."[168] He recognizes that the naive positivism of science seems to collapse so that there appear new starting points for a mutual understanding and a new cooperation of theological and scientific thinking.[169] We are confronted with a foundational problem of science and theology and are also forced to develop an ethos for the scientific technological conquest of the world. Yet Moltmann does not spell out how this cooperation between theology and the sciences is supposed to work. Rather he points out that in our modern industrial society, nature is being exhausted and natural resources are being relentlessly exploited, which destroys the natural foundations of life.[170] Technologies and the natural sciences are developed through certain human interests and are not value-free.[171] While in premodern cultures complex systems of equilibria were developed which regulated the relationship between humanity and nature and also the relationship of people among each other and to the gods, Western civilization furthered a one-sided development of growth and conquest.

As remedy for the ecological dilemma in which humanity is destroying the environment and itself in the name of progress, Moltmann points to the biblical Sabbath. "Sabbath rules are God's ecological strategy to protect the life which God has made. With its rest and its rhythm of time, the sabbath is also the strategy which will take us out of ecological crisis and after one-sided progresses at the expense of others shows us the values of abiding equilibrium and accord with nature."[172] "It would help us if we again recognized the wisdom of God in the laws and conditions of the system earth and in our own psychosomatic constitution. Living in accordance with God means living truly human lives, but we can live in accordance with God only if we also live in accordance with nature, in and with which we were made and through which God speaks to us."[173]

Moltmann also points out the necessity of cultivating Sunday as a

---

168. Jürgen Moltmann, "Theology in the World of Modern Science" (1966), in *Hope and Planning*, trans. Margaret Clarkson (New York: Harper & Row, 1971), 204.

169. Cf. Moltmann, "Theology," 208.

170. Cf. Jürgen Moltmann, *Creating a Just Future: The Politics of Peace and the Ethics of Creation in a Threatened World*, trans. John Bowden (London: SCM Press, 1989), 51.

171. Cf. for the following, Moltmann, *Creating a Just Future*, 52f.

172. Moltmann, *Creating a Just Future*, 66.

173. Moltmann, *Creating a Just Future*, 80.

day of rest and not turning it into an additional workday. Sabbath and Sunday belong to God. As humans we are created in the image of God and not as slaves of work and consumption. One should ask, however, whether the solution is as simple as Moltmann suggests. Automation has been and is needed in response to the steady reduction of the workweek. A further consequence has been that fewer people are needed for the same tasks. Furthermore, work time is used more intensively so that workers often need to take disability leave or retire earlier for health reasons.

When Moltmann considers creation, ecological problems are again at the forefront. Yet he does not think that an ecological catastrophe is unavoidable, because it was furthered through a questionable interpretation of creation. One interpretation of the doctrine of creation is that with the creation of the world everything was already given because creation was complete and very good and at the end there will be a restitution of creation. "The history between creation and redemption is then primarily the history of the Fall. It cannot bring anything new, except the increasing deterioration and aging of the earth."[174] If we regard, however, eschatology not from the vantage point of creation but vice versa, that is, creation from eschatology, then creation does not lie ahead of us like a closed system. It is, rather, open and will not necessarily change for the worse. Therefore Moltmann attempts to develop his doctrine of creation from an eschatological or messianic perspective.

Again he starts with the scenario of the ecological crisis, because for centuries one tried to understand God's creation as nature and to use it according to the laws of nature. Today, however, it is essential to understand the knowable, controllable, and usable nature as God's creation and to respect it accordingly.[175] In the history of the theological doctrine of creation Moltmann distinguishes three stages (33f.). In the first stage the biblical traditions and the worldview of antiquity were fused into a religious cosmology. In a second stage the natural sciences emancipated themselves from this cosmology, while theology detached its doctrine of creation from this cosmology and confined itself to personal

---

174. Jürgen Moltmann, "Creation as an Open System" (1976), in *The Future of Creation: Collected Essays*, trans. Margaret Kohl (Philadelphia: Fortress, 1979), 116.

175. Cf. Moltmann, *God in Creation*, 21. The parenthetical numbers in the following text refer to pages in *God in Creation*.

faith in creation. Only today, in the third stage, under the pressure of the ecological crisis and in the search for new directions, have theology and the natural sciences found each other again so that humanity and nature can survive on this earth. While the natural sciences show us how creation can be understood as nature, theology explains nature as God's creation. God's creation is neither divinized nor demonized, but understood as world. It is created by God and has no necessary existence. It is contingent.

From the point of view of a biblically understood Christology, we cannot regard the present state of the world purely as divine creation and as very good. We must rather emphasize that the longing and anxious waiting of creation indicate that it is open to the future of the kingdom of God, while it is presently subjected to futility. Moltmann argues that a biblical doctrine of creation cannot simply start with Genesis 1:2. "The starting point for a *Christian* doctrine of creation can only be an interpretation of the biblical creation narratives in the light of the gospel of Christ" (53f.). In the biblical traditions of the Old and New Testaments, the experience of the world as creation is determined by the belief in the self-disclosure of the creative God in the history of Israel. Since he has disclosed himself as creator, sustainer, and savior of Israel, God is also recognized as the creator of the world. In saying this, Moltmann relates salvation history and the experience of creation to each other. The experience of creation shows that the God who has made a covenant with Israel is the Lord and creator of the whole world and the whole universe. All human beings and all nations are included in the creative process which was experienced by Israel and was the basis for its hope. "Creation is the universal horizon of Israel's special experience of God in history" (54). Creation and salvation cannot be separated from each other, because creation is the universal horizon in which salvation is made possible. Conversely, salvation cannot be partially restricted to individuals or to a people; instead in creation it assumes cosmic and universal, but not necessarily universalistic, features. Moltmann, however, notes that the creation narratives do not yet yield a Christian doctrine of creation, since the messianic orientation is not sufficiently thematized in them.

Moltmann is correct when he emphasizes the ecological dimension of a Christian doctrine of creation and shows that the doctrine should not only contribute to a new intellectual self-understanding but must also lead to a new orientation of our very existence. Yet it remains to be seen if, biblically speaking, we can really talk about a messianic doctrine of creation. This can only be shown through the subsequent development of a Christian doctrine of creation.

We could review many other theologians who are explicitly concerned with the issues of the natural sciences in developing a Christian doctrine of creation, such as John Polkinghorne and Arthur Peacocke in Great Britain, Ian Barbour in the USA, or Günter Altner and Jürgen Hübner in Germany. Yet this would only lead to different accents but not to essentially new perspectives. After having slowly reached the present, we must now unfold the Christian doctrine of creation as far as it is possible given our scientific knowledge and the biblical witness. Yet before doing so, we should, for the sake of completeness, also briefly mention creationism.

Excursus: Creationism

Creationism advances a "creation science" which vehemently rejects any evolutionary thinking. Its main argument is the same as what Charles Hodge advanced against Darwinism in the nineteenth century in *What Is Darwinism?* Hodge distinguished between the ideas of Darwin and Darwinism and claimed the latter would lead straight to atheism. Creationism gained international renown through the Scopes trial, though it did little to advance the creationist cause. On the contrary, it discredited the fledgling "science."[1]

Since the 1960s there has been a reawakening of creationism, and it has gained increasing momentum. In the summer of 1980 the first issue of *Creation/Evolution* appeared. It was the first journal to be exclusively dedicated to the advance of creationist ideas. This journal was supplemented by the *Creation/Evolution Newsletter*.[2] In the USA attempts were made to introduce creationism as an official doctrine in high school curricula with the argument that creationism is a science which would merit the same attention as evolutionary science. Otherwise the freedom of information of creationistic students would be limited. This shocked the National Academy of Sciences and the National Association of Biology Teachers in the USA so much that in October 1981 they convened special meetings in Washington because they had recognized that creationism could no longer be discarded as an absurdity like the

1. Cf. to this point Ronald L. Numbers, "The Creationists," in *God and Nature: Historical Essays on the Encounter between Christianity and Science*, ed. David C. Lindberg and Ronald L. Numbers (Berkeley: University of California Press, 1986), 402.

2. Regarding the phenomenon of creationism, see the detailed investigation by Ronald L. Numbers, *The Creationists* (New York: Alfred A. Knopf, 1992), esp. 320f.

theories of the "Flat Earth Society." A decision of the American Supreme Court in 1987 left open the possibility that on a voluntary basis creationistic science could be taught. Many teachers pursued this course of action to evade a confrontation with creationists. According to a survey, 30 to 69 percent of teachers in public schools decided to include creationism in their curricula.

In contrast to the antievolution crusades of the 1920s, the recent creationist movement has not been limited to North America but covers practically the whole earth. In Europe the pharmacologist Arthur E. Wilder-Smith (1915-95), a native British citizen, has been so influential that often he is regarded as "Europe's leading creationist scientist."[3] In his youth he joined the fundamentalistic Plymouth Brethren. He obtained three doctorates, one in physical organic chemistry, another in chemotherapy, and a third in pharmacology. He has lectured in Germany, Switzerland, the USA, Norway, and Turkey. Many of his books are published in several languages and have had high sales figures. In Germany a creationistic society was established in the late 1970s, and a monthly publication, *Factum,* was published. Especially engaged in creationism was a group of people who left the Karl Heim Society and established their own research institute, also publishing a series of books at the Hänssler Publishing House. In rapid sequence numerous books have appeared which, for the most part, sought to discredit the theory of evolution.[4]

The concern of the creationists is understandable. In a world which is ever more subject to rational scrutiny and in which God is pushed to the corners, the meaning and direction of life, of society, and of individuals becomes haphazard. Creationists attempt to introduce a radical change on a biblical basis. Problematic, however, is the fact that they want to introduce creationism as a scientific theory and as a credible scientific alternative to the theory of evolution. They use a literalistic interpretation of the Bible which does not do justice to the intent of the creation accounts. Moreover, they attempt to mold the creation accounts into the frame of contemporary scientific thought and bring them into a context in which they do not belong unless one disregards the historical background of the accounts. By turning the creation ac-

---

3. Numbers, *The Creationists,* 334.
4. Cf. n. 84 above.

counts into scientific reports, no attempt is made to break through the exclusivistic scientific rational view of life and of our environment. To the contrary, such a view is strengthened. Creationism wants to confront the scientific attitude of evolutionism on its own turf. It takes on the characteristics of evolutionism and becomes as exclusive and ideological as a reductionistic evolutionism in which God does not appear in its evolutionary worldview, a worldview in which God does not exist and does not participate in the events of the world.

The American Scientific Affiliation, by contrast, is much more pragmatic and credible in its approach. Though evangelical in persuasion, it concedes an evolutionary understanding of the world as a working hypothesis.[5] At the same time it warns against reductionistic evolutionism. Therefore the American Scientific Affiliation can affirm the divine inspiration, trustworthiness, and authority of the Bible for faith and conduct as well as investigating any questions which relate to the Christian faith and the natural sciences without preconceived notions.

5. Cf. "Statement of Faith" of the American Scientific Affiliation, printed in *Perspectives on Science and Christian Faith* (journal of the American Scientific Affiliation).

5. Developing a Christian Understanding of Creation

The Christian understanding of creation is first of all informed by the biblical witness to God's creative activity. Then it endeavors to express its content with metaphors that are understandable today. To understand the Christian doctrine of creation we must therefore look to the Scriptures and discern the biblical view of creation. This also means that we take an especially close look at the first two chapters of Genesis, since they focus on creation and may well set the tone for everything else said about creation in the Bible.

a. The Understanding of Creation according to the Biblical Witness

In the Old Testament the term *bara* (to create) has special theological significance. It is "used to express clearly the incomparability of the creative work of God in contrast to all secondary products and likenesses from already existing material by man."[1] *Bara* is used exclusively for divine activity. It is not without significance that this term is used primarily in texts which date from the exilic or postexilic period. "Therefore, it can be regarded as certain that *bara* was introduced into OT literature as a theological idea for the first time in the exilic pe-

---

1. Karl Heinz Bernhardt, "bara," in *Theological Dictionary of the Old Testament,* ed. G. J. Botterweck and H. Ringgren, 2:246.

riod."[2] In Genesis 1:27 *bara* is used three times in connection with the creation of humanity. In Deutero-Isaiah the verb *bara* is associated with salvation. It refers to the activity of God in both the distant past and the immediate future. Therefore we read, for instance, in Isaiah 48:7: "They are created now, not long ago." This new creative activity extends also to the forces of nature, so that barren land is turned into a fertile and wooded area (cf. 41:18ff., where *bara* is used in v. 20). In 45:8 we even read that God creates *(bara)* the conditions under which salvation can originate.

The relatively late occurrence of a term for "creating" as God's own prerogative and the frequent connection of creation with salvation made the German Old Testament scholar Gerhard von Rad (1901-71) assert that faith in God's creative activity is not a doctrine which is a topic in its own right. It is always related to something else and subordinated to the interests and content of the doctrine of salvation. This soteriological interpretation of God's creative activity is, according to von Rad, "the most primitive expression of Yahwistic belief concerning Yahweh as Creator of the world."[3] Even the creative narration in Genesis 1, according to von Rad, is completely motivated by God's interest in salvation. The Yahwist faith in creation never enjoyed the status of an independent doctrine. Yet von Rad concedes that this need not imply that the belief in God's creative activity has a late origin. Indeed, there is widespread agreement that, while Genesis 1 is postexilic, the Yahwist creation narrative in Genesis 2 is much older and belongs to the period of the early monarchy.[4]

In a later publication, *Wisdom in Israel* (1970), von Rad pointed out that the Wisdom tradition in Israel picked up the theme of creation quite early. This tradition is part of the oldest layers of the Old Testament and provides a very different understanding of God and the world from other Old Testament traditions but still witnesses to the same faith in God. According to von Rad, the Wisdom tradition turns to nature instead of to history and is universal instead of particularistic in its view.

2. Bernhardt, 2:245.

3. Gerhard von Rad, "The Theological Problem of the Old Testament Doctrine of Creation" (1936), in *The Problem of the Hexateuch and Other Essays,* trans. E. W. Trueman Dicken (Edinburgh: Oliver & Boyd, 1966), 138.

4. So Brevard S. Childs, *Biblical Theology of the Old and New Testaments: Theological Reflection on the Christian Bible* (Minneapolis: Fortress, 1993), 107.

In his exegesis of three major hymns (Job 28; Prov. 8; Sir. 24), von Rad shows that only in Israel can wisdom learn from "the self-disclosure of creation." Von Rad states: "the doctrine of the primeval revelation with its distinctive element — namely the address to men — stands, therefore, on a genuinely Israelite basis."[5] We encounter here a doctrine of revelation "which happens to men not through a specific, irreversible sign of salvation in history, but which, rather, emanates from the power of order which is held to be self-sufficient." The conviction expressed here is that the world is not silent but proclaims God as creator (cf. Ps. 145:10f.) — which would justify Pannenberg's challenge to scientists to deepen and broaden the implications drawn from their scientific investigations. Assertions concerning creation are found outside the two creation narratives especially in Israel's hymns, such as in Psalms 8, 100, 104, and 144. But we must also pay attention to Deutero-Isaiah since there, as stated above, a close connection is made between creation and salvation.

To obtain a balanced view of how creation is valued, we must also turn to the New Testament. There the verb *ktizo* is used exclusively to signify a divine creative activity.[6] The New Testament writers are clear on the point that God created the world, the heavens, the earth, and everything on it. It goes without saying that there is no preexistent matter (cf. Rom. 4:17), because everything is created (Eph. 3:9). There is also no emanation of matter, meaning an ushering forth of the created from the Godhead. Everything which is created and limited is also subjected to temporality and is perishable. This also pertains to humanity, which is related to its creator like clay to a potter (Rom. 9:20ff.). Humanity is totally in God's hands and depends on God's grace. But if someone is in Christ, "there is a new creation: everything old has passed away; see, everything has become new!" (2 Cor. 5:17). There is an indication of a new creation in which the whole of creation will someday be included and be released from its bondage to decay (Rom. 8:21). Creation in the beginning, therefore, has to be connected with the new creation since creation in its present state does not correspond to the glory of God. Cre-

5. So Gerhard von Rad, *Wisdom in Israel* (Nashville: Abingdon, 1972), 175, for this and the following quotation.
6. So Werner Foerster, "ktizo," in *Theological Dictionary of the New Testament*, ed. G. Kittel and G. Friedrich, 3:1028.

ation therefore must also be understood in its historical context. It is not a singular fact which occurred in the beginning, but it contains a dynamic element.

b. The Biblical Creation Narratives

Biblical theology usually distinguishes between two creation narratives: the Priestly narrative in Genesis 1:1–2:4a and the Yahwist narrative in Gen 2:4b-24. Right at the beginning of the Priestly narrative we read the important theological confession: "In the beginning when God created the heavens and the earth. . . ." The subsequent sevenfold partition of the narrative is supposed to show that everything in the cosmos owes its existence to God's power and might.[7] The idea has often been advanced that God's creative activity culminates in the creation of humanity. Yet one should not overlook that at the end of creation there is no mention of humanity but of the seventh day as a day of rest. The postexilic character of this narrative as it now stands becomes at once evident, as we will point out, and therefore it may have originated in the sixth or fifth century B.C. Yet individual traces of this narrative seem to reach far back into Israel's history.

In the narrative God is described as a mighty or divine wind that moves above a dark, undifferentiated watery matter. Through the power of his word God creates the cosmos out of this void. In antiquity this kind of creative activity was not unknown in the Palestinian and Mesopotamian Near East. While in many creation myths the world was understood as an emanation from the Godhead, an ushering forth of God's essence, in the Genesis account God creates the world by his word.[8] This illustrates that the Creator creates the world without effort, and at the same time it points to an ontological difference between the Creator and the created. Nothing divine is contained in that which we call creation. It is created, and therein consists its actual dignity. It is also significant that God did not say regarding vegetation: "Let there be

---

7. For the following, cf. Robert A. Oden, Jr., "Cosmogony, Cosmology," in *The Anchor Bible Dictionary* (New York: Doubleday, 1992), 1:1166.

8. For the following, cf. Gerhard von Rad, "The Biblical Story of Creation," in *God at Work,* trans. John H. Marks (Nashville: Abingdon, 1980), 100.

vegetation!" but "Let the earth put forth vegetation" (Gen. 1:11). Similarly we hear in Genesis 1:20: "Let the waters bring forth swarms of living creatures." The earth and the waters were called upon to be actively involved in God's creative activity.

It might be confusing for us that, at first, light is separated from darkness (Gen. 1:4) and only much later the sun, the moon, and the stars are created (1:14ff.). Though there are certain analogies between this creation narrative and that which most scientists hold today to be the likely course of the evolution of the world and of life, modern scientific knowledge also points out the big differences between the two. This leads us to the conclusion that Genesis 1 portrays a worldview and a knowledge of nature which is quite different from ours. We have acquired immense knowledge about nature since this narrative was first written down. Therefore Gerhard von Rad states: "Christians have had to learn, amid severe shocks to their faith, that this view of the world is quite antiquated."9 Von Rad would not find it helpful to modernize the creation narrative by replacing an outdated view of nature with a more modern one. By so doing we would tear apart what goes hand in hand here: theological and scientific knowledge. "Theology found in the science of that time an instrument it could use unhesitatingly to unfold the content of faith."

Modernizing the Priestly creation narrative would cause its theological intention to be totally lost. The thrust of this creation narrative is not to show that a symbiosis of theology and science does exist, though it expresses its intention in this symbiosis, but its intention is the same as that of the psalmist who states in Psalm 33:8f.:

> Let all the earth fear the LORD;
>   let all the inhabitants of the world stand in awe of him.
> For he spoke, and it came to be;
>   he commanded, and it stood firm.

From beginning to end do we notice an apologetic tendency in Genesis 1: against all other gods and powers dominant at that time the God of Israel is confessed as creator and Lord of the whole world. Right at the beginning God fearlessly towers over Tiamat, the life-threatening goddess of the seawater. In Babylonian creation mythology the abysmal

---

9. Von Rad, "Biblical Story of Creation," 105f., for this and the following quote.

depth, the *Tiamat,* as she was called and feared in Babylon, was slain by the youthful god Marduk after a fierce battle. But in Genesis 1 God simply commanded her to recede to the depth so that the dry land might appear. In a similarly sovereign act God "glued" the stars onto the sky. Even sun and moon were created by him to rule over day and night respectively. Such assertions were made in Babylon, a country in which the moon was venerated as deity and the stars were believed to determine the life of the people. In this environment, in which the Israelites lived in exile, their religious leaders dared to reject the power and influence of these forces of nature. The Priestly creation account was not composed by a mighty and powerful nation which in some kind of self-glorification would have asserted its God as the only creator and ruler of the world. It was the opposite. In captivity and exile this daring and yet congruent affirmation of God as creator and ruler of the world was first made. Any kind of modernization to make it scientifically more amenable would bring to naught the apologetic intent of this narrative. Even if its worldview is outdated, though we can also detect modern-looking features, the theological intent of the Priestly creation account, still valid today, does not allow nor need any modernization.

The Yahwist narrative, which begins in 2:4b, does not present an account of the creation of the world. Strictly speaking it presents only an account of the creation of humanity. This event dates back to the earliest history and presupposes the actual creation of the world. The only reminder of such a creation is seen in the fact that God had not yet caused it to rain upon the earth. Of course, it is Yahweh who gives the rain and thereby the possibility for things to grow and thrive. The threatening seawater, of which the Israelites did not learn until the Babylonian exile, is not mentioned. But with a few strokes a very different and earlier environment is sketched out: "It is a world where people cultivate the soil, surrounded by steppe and desert, where life depends on the rain that gives growth to the shrubs of the steppe and to the seed of the cultivated land. It is a world which corresponds to that of the Palestinian farmer."[10] The immediate environment of humanity is described. It is a matter of fact that God has created everything and that he gives the water which allows life to thrive. In contrast to the Priestly account

10. So Claus Westermann, *Genesis 1–11: A Commentary,* trans. John J. Scullion (Minneapolis: Augsburg, 1984), 1:200, in his exegesis of Gen. 2:5.

in which humanity is placed at the end of the creative process, here everything is created around humanity, so that it can feel at home in God's creation.

We must also include in our reflections the so-called Noachic blessing in Genesis 8:22. After the flood narrative "there is a creation statement which takes its place between the creation of humans in Gen 2 and that of the world in Gen 1."[11] For the first time in human history the cosmic events and their temporality are understood as a unity. Life has its own existence and an order which is appropriate to it. Not forever, not in eternity, but as long as the earth exists will there be seedtime and harvest, cold and heat, summer and winter, day and night. Life needs the cyclical changes of day and night for human food to grow. The life of plants and animals is shaped by the rhythm between seedtime and harvest, summer and winter. Through God's decision this rhythm, which makes life possible, is granted and sanctioned. It is no longer the garden but the world itself which becomes the arena of life, sanctioned by God. Yet before we assess life in this world we must return to the question of a creation at the beginning.

c. Creation in the Beginning *(Creatio ex Nihilo)*

The opening sentence of the Priestly creation narrative should give us an answer to the question of whether God created the world out of nothingness. But the opening line, "In the beginning when God created the heavens and the earth," is not so self-explanatory as we may generally assume.[12] At the most, this assertion indicates the flow of thought for subsequent statements. It contains nothing more than the conviction that God created the world.[13] Whether this also implies that God created a primeval chaos which, so to speak, is presupposed by the subsequent creation, which is ordered, is a consideration the Priestly writer did not yet make.

The first explicit assertion of a creation out of nothingness is not made until 2 Maccabees 7:28, where we read: "I beg you, my child, to

11. Westermann, 1:457, in his exegesis of Gen. 8:22.
12. Cf. Westermann, 1:93, in his exegesis of Gen. 1:1.
13. For the following, cf. Westermann, 1:109ff., in his exegesis of Gen. 1:2.

look at the heaven and the earth and see everything that is in them, and recognize that God did not make them out of things that existed." When Greek patterns of thinking and Greek conceptuality were received into Judaism, such a statement made sense. But in earlier times one did not reflect in Israel about being or nonbeing. Even for us today it is difficult, if not impossible, to think of complete nothingness or to "imagine" such nothingness. One talks about it either philosophically or metaphorically. Exactly the latter occurs in Genesis 1: "From the starting point of the actual world of his own experience, he [the Priestly writer] describes a state of affairs which is the very antithesis of it, in which things are *not* ordered and clearly distinct from each other, indeed are simply not there at all."[14] This antithesis to our present world is described by the Priestly writer as chaos, as something not chronologically but rather logically prior to the world. In a similar way the ground on which nothing grew in Genesis 2 functions as the antithetical starting point to creation. We have been educated by Greek thought always to put a nonbeing prior to an initial singularity and to emphasize the creation out of nothingness in order to express the transition from no matter to some matter. The Priestly narrative conveys the same intention with the transition from a nonordered to an ordered state.

For us the notion of a creation out of nothingness means that God acts without preconditions. There is nothing prior to the creation which limits God's sovereignty over against the world. This means that a model of creation as expressed by process thought is insufficient for the biblical intention because God is more than a creative participant in cosmic events. Ian Barbour muses: "God is like a teacher, leader, or parent. But God also provides the basic structures and the novel possibilities for all other members of the community. God alone is omniscient and everlasting, perfect in wisdom and love, and thus very different from all other participants."[15] This community thinking contradicts the biblical affirmation of creation. Nothing can influence the creative power of God and nothing can resist it. For God creation was neither a necessity nor an accident. It occurred because God wanted it so.

14. Henricus Renckens, *Israel's Concept of the Beginning: The Theology of Genesis 1–3*, trans. Charles Napier (New York: Herder and Herder, 1964), 51, in his interesting reflections.

15. Ian G. Barbour, *Religion in an Age of Science* (San Francisco: Harper, 1990), 269, who prefers the process model over all other models.

Having said this, we should not too quickly identify the temporal beginning of the universe in a creation out of nothingness with the cosmic big bang. Such a cosmic singularity, from which the universe expands, need not necessarily presuppose that there was nothing prior to it.[16] At the beginning we still have an initial singularity. Yet this expansion from a starting point can indicate that the world has been created, that it has not always been the way it is presently known to us. The first event, which gave space, time, and matter its measurability, was, as the Genesis narratives assert, a creative act of God through which something came into being. If we assume such a starting point, such an initial singularity, then all scientific theories about the evolution of the universe and of life in it have their appropriate starting point.

If we want to express the content of the creation narratives with contemporary conceptuality, then we could follow the suggestion of the nuclear physicist and theologian William Pollard and say: "When God began to create the universe, there was nothing; no space, no time, no matter; nothing at all. Then there was a black hole, shrouded in impenetrable mystery. Out of this black hole came an expanding space-time universe, initially a vast fireball filled with light, neutrons, protons, electrons, and anti-neutrons, all governed by definite physical laws."[17] Such modernizations of the creation narrative are subject to continuous revisions because of the steady increase in scientific knowledge. Yet the fundamental theological intention of these assertions remains the same and is congruent with the basic convictions that we notice in both biblical creation narratives: God is the creator of everything that is and without him there would be nothing.

Here we notice the difference between the voluntary self-limitation of science, which does not venture any assertions beyond the space-time or matter-energy continuum, and theology, which deepens these scientific details and enlarges them in asserting to whom this continuum owes its existence and structure. While science describes certain states and secular interpretations of the world talk about chance and necessity, theology understands the world in a theocentric way. The world is seen

16. See also the legitimate warning by Pannenberg, "Theological Appropriation of Scientific Understandings: Response to Hefner, Wicken, Eaves, and Tipler," *Zygon* 14 (1989): 264f.

17. William Pollard, Oak Ridge, Tenn., in a letter to the author dated October 16, 1976.

as the expression of God's will and power. Theology brings God and the world into a relationship and therefore talks about creation and the created, while science in its self-imposed limitation attempts to understand the world without God and therefore talks about nature. One might wonder, however, whether this self-imposed limitation which is scientifically necessary can still be asserted with regard to the human situation. Here the problematic nature of this reduction becomes especially noticeable when we perceive humanity both as a creature and as someone who is involved in the creative process.

d. Humanity as Creature and Cocreator

In the dialogue with the natural sciences the Lutheran systematician Philip Hefner describes humanity as "created co-creator."[18] The term "co-creator" corresponds to the freedom of humanity which should be understood neither in the liberal sense as self-determination nor in the Marxist sense as the ability to make and shape the world. Freedom is rather the condition of existence in which humanity is unavoidably confronted with the necessity to decide and to construct narratives which provide a context for its decisions and thereby justify them. Hefner asks whether the concept of cocreator is not too arrogant and not too narrowly anthropocentrically conceived. He wonders if it would not strengthen the egocentric and self-seeking behavior of humanity which confronts us in its ecological ramifications.[19] Humanity, by virtue of its genetic programming and neurobiological possibilities, has been endowed with unique capabilities. Hefner is persuaded that human history demonstrates that humanity has used these capabilities to shape the earth more and more to its own liking and thereby threaten it. Yet God's intention for humanity was the exact opposite: human conduct should have a positive effect on creation. Both aspects, the threatening and the positive, show for Hefner that humanity is not only a creature but also a cocreator. The idea of the created cocreator can be discerned in the creation narratives also.

18. For the following, see Philip Hefner, *The Human Factor: Evolution, Culture, and Religion,* with a preface by Arthur Peacocke (Minneapolis: Fortress, 1993), 38.
19. Hefner, *The Human Factor,* 236f.

In the Priestly narrative four verses talk about humanity and its creation. In Genesis 1:27 we read three times that God created humanity, as if to emphasize that humans are definitely God's creatures, no more and no less. They are very closely associated with animals, because there is no special day of creation reserved for humans. Humans are created on the same day as the land animals. Like the fish of the seas and the birds of the air, they are blessed and given the command to be fruitful and multiply. They are even supposed to share the same food with land animals. The Hebrew word that is used to denote a human being, *adam,* is a collective word, actually meaning humankind. Only later does it become a proper name for the first human being, Adam. To ensure that it is not misunderstood as depicting some kind of archetypal primal androgynous being, the immediate qualification is added that God created "them, male and female." Sexual differentiation is something given with creation. A human being that does not occur either in a male or a female form is an impossibility, since humanity is only present in one of these two forms. "Every theoretical and institutional separation of man and woman, every deliberate detachment of male from female, can endanger the very existence of humanity as determined by creation."[20]

The Yahwist creation narrative shows too that the creation of humanity is exclusively the work of God. The Yahwist account reads: "Then the LORD God formed man from the dust of the ground, and breathed into his nostrils the breath of life; and the man became a living being" (Gen. 2:7). Gerhard von Rad called this verse "a locus classicus [a classical assertion] of Old Testament anthropology."[21] There is nothing divine in humanity, since it is closely associated with matter. Though we might expect that the creation narrative makes a distinction between body and soul, such a distinction is foreign to biblical thinking. Humanity is perceived as a unity. A human *(adam)* is taken from earth *(adamah).* Therefore we read toward the end of the story of the fall,

> You are dust,
> and to dust you shall return. (Gen 3:19)

20. Westermann, 1:160, in his exegesis of Gen. 1:27.

21. Gerhard von Rad, *Genesis: A Commentary,* trans. John H. Marks (Philadelphia: Westminster, 1961), 75, in his exegesis of Gen. 2:7.

Humans return back to nature, from which they were taken. But earth does not simply yield humans, as the Priestly account witnesses. There the Priestly writer reports that the earth brought forth living creatures of every kind. There is no natural necessity for humans to emerge. God himself gives his breath into a human, and so there emerges a living human being. Human life is not taken for granted, it is a gift of God. We receive our human existence by participating in God's life-giving spirit. When many people question today whether humanity has a chance of surviving the ecological crisis, this reference to the gift character of human life takes on renewed significance. Humanity has no right of its own to survive. Life and the future which is necessary for life to unfold are God's gifts.

The creator of all of humanity is at the same time the creator of each individual human being (cf. Isa. 17:7). The psalmist writes of this creator:

> For it was you who formed my inward parts;
> > you knit me together in my mother's womb.
> I praise you, for I am fearfully and wonderfully made.
> > Wonderful are your works. (Ps. 139:13f.)

Similarly we read in Job:

> Your hands fashioned and made me;
> > and now you turn and destroy me.
> Remember that you fashioned me like clay;
> > and will you turn me to dust again?
> Did you not pour me out like milk
> > and curdle me like cheese?
> You clothed me with skin and flesh,
> > and knit me together with bones and sinews.
> You have granted me life and steadfast love,
> > and your care has preserved my spirit. (Job 10:8-12)

Though the writer of the book of Job describes the procreative process, that the semen is injected into the female organism and a solid embryonic body is formed ("Did you not pour me out like milk / and curdle me like cheese?"), this whole process is, at the same time, understood as a work solely wrought by God. It is beyond the imagination of the Old Testament writers that nature should do its part and then God would do the rest, or that the natural processes would involve some kind of automatism.

Martin Luther (1483-1546) captured well the biblical understanding of the dependency of human life on God when he said that God "could give children without using men and women. But He does not want to do this. Instead, He joins man and woman so that it appears to be the work of man and woman, and yet He does it under the cover of such masks."[22] Each human being is totally God's work regardless of his or her line of descent. The question of whether life evolved "naturally" (unaided) from inanimate matter, though it may be scientifically significant, is theologically secondary. Both inanimate nature and living creatures are God's creation. Whenever and wherever human life appears it is a gift of God.

After this brief survey of additional assertions concerning creation, we must once more return to the Yahwist creation narrative. Humanity's creaturely position is not only indicated by the Yahwist when he mentions that humans are taken from dust and created through the life-giving spirit of God. Their reddish brown skin reminded the Israelites of the reddish brown dust of the earth. Moreover, animals are seen in close relationship with humans. Like humans they are created from dust, and they are even introduced as possible helpers for humans (Gen. 2:18). Finally, the first human being was supposed to name the animals. In these names they receive the designation the first human attributes to them.[23] The naming of the animals is no magic act through which a human might obtain power over the animals, but rather, with these names, this human puts them into a place in his world. They are part of his world and belong to him. If today more and more species of animals are destined to extinction, the human world is also impoverished. Animals belong to humans and are part of the richness of the human world.

When we turn to the creation of the woman, we notice that God's creative activity does not tolerate any spectators. Adam falls into a deep sleep when Eve, the mother of all living beings, is created. We are only allowed to view God's creation retrospectively. "The creation of woman

---

22. Martin Luther in his exegesis of Ps. 147:13, in *Luther's Works*, 55 vols. (St. Louis: Concordia; Philadelphia: Fortress, 1955-86), 14:114; hereafter cited as *LW*.

23. For the following cf. Westermann, 1:228, in his exegesis of Gen. 2:19.

from the rib of the man should not be understood as a description of an actual event accessible to us."[24] Through the creative event itself, the mutuality of man and woman is explained. The woman is from the same "material" as the man, from human stuff. Man *(ish)* notices this immediately and calls her "woman" *(ishah)*, similar to the old English meaning of "woman" (female human).

In the beginning of the narrative we read that God created "a helper as his partner." This was often interpreted to mean that the woman was to help the man in agriculture. Feminist theology has rightly corrected this exegesis which demeans the woman as the handmaiden of the man. As the Yahwist creation narrative shows, such a devaluing of the woman is not intended, but there is expressed a personal correspondence between man and woman. With the creation of the woman the creative process of humanity is concluded. Therefore one could even regard the woman as the pinnacle of God's creation. Yet the woman does not stand alone. Only now can there be "the personal community of man and woman in the broadest sense — bodily and spiritual community, mutual help and understanding, joy and contentment in each other."[25] When we read in the concluding sentence that a man leaves his father and mother and clings to his wife, this is to show again the strong relationship between man and woman which is sustained through the love between them and against which even the bond with the parental home cannot supervene.

In our short review we notice that the creation narratives disagree in details, but there is a fundamental agreement in their intentions. Humans are seen as creatures, placed in immediate vicinity of animals, but created in distinction to the animal world and in a basic twofold appearance, namely, of men and women. Yet we have still left out an important aspect of being human. In the Priestly narrative we read that

> God created humankind in his image,
>> in the image of God he created them. (Gen. 1:27)

It is important that we try to discern what it means to be created in God's image.

In Genesis 1:26ff. we read: "Then God said, 'Let us make human-

24. Westermann, 1:230, in his exegesis of Gen. 2:21f.
25. So to the point Westermann, 1:232, in his exegesis of Gen. 2:23.

kind in our image, according to our likeness; and let them have dominion over the fish of the sea, and over the birds of the air, and over the
cattle, and over all the wild animals of the earth, and over every creep-
ing thing that creeps upon the earth.'" This passage has been the cause
of many heated discussions, not only because of the assertion that humans are created in God's image, but also because of the command that
humanity is to have dominion over the earth. Are humans thereby not
put into immediate proximity to God? And is humanity allowed to subject everything on earth to its advantage and use it according to its own
ideas?

The American historian of science Lynn White, Jr. (b. 1907), called
Christianity the most anthropocentric religion the world has seen, because nature's sole purpose for existence is to serve us.[26] In a similar
manner the German critical essayist Carl Amery (b. 1922) wrote of the
merciless consequences of Christianity.[27] Already in the nineteenth
century Ludwig Feuerbach asserted: "Separation from the world, from
matter, from the life of the species, is therefore the essential aim of
Christianity."[28] In his insightful study, Paul Santmire, however, showed
that the theological tradition of the West is neither ecologically bank-
rupt, as many assert, nor does it simply suggest simplistic solutions for
averting our ecological crisis. Santmire contends that the Christian faith
contains starting points "for a rich theology of nature."[29] At the same
time he also admits that in Western theology humanity always has to
rise far beyond nature to enter into communion with God.[30] The subsequent rift between humanity and nature can, of course, become problematic for both.

If one looks carefully at Genesis 1:26ff., one can easily discern the
special position of humanity. Already in the initial move, "Then God
said, 'Let us make humankind,'" we can discern a special decision of

26. Lynn White, Jr., "The Historical Roots of Our Ecologic Crisis," first published
in *Science* 155 (March 10, 1967): 1203-7, and since then often reprinted.

27. Carl Amery, *Das Ende der Vorsehung: Die gnadenlosen Folgen des Christentums*
(Reinbek bei Hamburg: Rowohlt, 1972).

28. Ludwig Feuerbach, *The Essence of Christianity*, trans. George Eliot, introduction
by Karl Barth, foreword by H. Richard Niebuhr (New York: Harper & Row, 1957), 161.

29. So H. Paul Santmire, *The Travail of Nature: The Ambiguous Ecological Promise of
Christian Theology* (Philadelphia: Fortress, 1985), 8.

30. Cf. Santmire, 188.

God which is missing with respect to the other acts of creation. Again we must take notice of the historical context, because this text was shaped by the experience of the Babylonian exile. According to Babylonian mythology, "people were created to minister to the gods."[31] But of this we hear nothing in Genesis. On the contrary, the purpose of humanity focuses on activities within the world, to have dominion over animals. "The creation of human beings introduces the possibility of a hierarchical order which is characteristic of 'being in the world.' The goal of the creation of humans is detached from the life of the gods and directed to the life of this world."[32] Now we can see the reason for the creation of humanity. They are not created as domestic servants of the gods, but have their own task here on earth.

Nevertheless, it is asserted that humanity was created in God's image. When the Hebrew language uses two different words to denote this, *demut* and *zelem,* this does not imply two different kinds of images, but the two words are to be understood as synonymous. Yet the later Latin tradition distinguished here between *imago* and *similitudo,* and asserted that the *imago* image was lost through the fall while the *similitudo* image remained. Since only man and woman were created in God's image, it is not surprising that humanity has an elevated position over against other creatures. The being which was created "in the image of God" separates humanity from all other creatures made by God. One should not speculate, however, on which parts or functions of humanity correspond especially well to being created in God's image. Such differentiating speculations are foreign to this text. We are simply reminded that the whole person is created in God's image.

A view of the cultural historical context can be enlightening.[33] In Mesopotamia and in Egypt images of the gods were often used in religious cultic life. The significance of an image did not lie in the fact that it would describe or actually depict the godhead, though this may not have been completely off the mark. What was decisive was that the image was the place at which the godhead was present and made manifest. The presence of the god and the blessing of his or her presence was me-

---

31. So Westermann, 1:157, in his exegesis of Gen. 1:26f.

32. Westermann, 1:159.

33. To the following, cf. Edward M. Curtis, "Image of God (OT)," in *The Anchor Bible Dictionary,* 3:390f.

diated through the image. In Egypt the pharaoh was even regarded as the earthly manifestation of the godhead and functioned analogously to the image of the god which was kept in the temple. In Mesopotamia the king was only considered for a short period the representation of the godhead. In Israel, according to Genesis 1, the image of God was not confined to the king but was extended to humans in general. Therefore, being created in the image of God consists in this: "that the human being rules over the rest of creation as king, governor, and God's representative on earth."[34]

According to Genesis 1:26ff., being created in the image of God does not imply a special ontological quality, but it is an assertion about the function of humanity. That is to say that humanity has been created and is called forth to rule over the rest of creation. This is also expressed in Psalm 8, where the psalmist writes:

> What are human beings that you are mindful of them,
>   mortals that you care for them?
> Yet you have made them a little lower than God,
>   and crowned them with glory and honor.
> You have given them dominion over the works of your hands;
>   you have put all things under their feet. (vv. 4ff.)

Being created in the image of God is not intended to deify or idolize humanity; it is also no license to exploit creation and to subjugate it to one's desires. Being created in God's image means rather to act in God's place, as his administrator and representative. This understanding is reinforced in the New Testament. The Pauline corpus is the main — almost exclusive — source of this understanding. In the New Testament, to be created in God's image means to be ethically shaped in conformity with God and to act in a manner for which God serves as the prototype. Such conduct can be derived from God's ways as they are transparent in Jesus of Nazareth (cf. Phil. 2:5; Rom. 15:5).

As humans we are supposed to represent God and to model our conduct according to God. This also illustrates the limits of human freedom. It is not a freedom to do whatever we want to do, but rather the freedom to live according to God's intentions. We should remember that every creative act of God was described as "very good." As God's

34. Jacob Jervell, "Bild Gottes I," in *Theologische Realenzyklopädie,* 6:492.

representatives we should maintain and preserve this "very good." We should direct and channel everything which stands against God's creation, which is destroyed, corrupted, and in bondage to decay, into the direction of this "very good." In so doing we would be truly creatures, but also cocreators because we would exercise our responsibility for creation so that it does not develop against its original intention or is used contrary to it. We are God's representatives and derive our self-understanding from the task of representing God. Burdening creation, however, through mindless procreation and pushing it beyond its carrying capacity (Gen. 1:28 says we should merely fill the earth, not overpopulate it!), and exploiting its natural resources so that subsequent generations are surrounded by garbage dumps, have nothing to do with representation but with egotism. Yet our experience of God as a loving and caring God should give us pointers to how we as his representatives should be experienced by others. Being created in the image of God contains authority but at the same time humility. Decisive action would be paired with caring, and loyalty in service to others with dignity.

Contrary to God's original intention, we could also in an emancipatory way use our own selves to define our existence. We would be the sole measure of all things and everything would have to serve us. We would be confronted not only with nature, but as individual human beings also with all other human beings. Our modern experience as solitary and isolated and individual human beings shows ever more clearly the precariousness of our existence if we no longer understand ourselves as being created in God's image. We abandon our responsibility of caring for and cultivating God's creation. Instead we regard it as our property while simultaneously ignoring its comprehensive context. Trying to gain the world for ourselves, we are on a certain path to losing it and ourselves. If a finite human being becomes the measure of everything, then it must absolutize itself and regard itself as infinite. This means it lives in contradiction to itself and to everything else. But God has not only created the world and humanity to relieve them from the burden of living contrary to their innermost being, but he has also given them a certain order through which he lovingly maintains them. This order should now be elucidated.

e. Divine Preservation of Creation
(General Providence)

Divine providence seems to have lost its persuasive power. Confronted with our world which becomes increasingly inhuman, the urgent question arises of whether anybody still cares about the fate of our world. As a consequence of our rapid technological development through which humanity threatens itself and the world in which it lives with extinction, many people are asking which direction to take and whether there is a direction that still makes sense or whether everything does not end in meaninglessness. Though we are yearning for the comforting assertion that the future does not confront us as an inescapable fate, but that there is meaning and fulfillment for our lives and for the cosmos in general, we cannot close our eyes to modern critical rationality. According to that thinking, each effect must have a cause which can be rationally explained. This would mean that divine providence and preservation of the world is hardly tenable today. William Pollard rightly says that "to speak of an event as an act of God, or to say that it happened because God willed that it should, seems a violation of the whole spirit of science."[35]

Not only our secular rationalism makes it difficult for us to talk about divine providence. Modern technology has given us ever greater possibilities of changing the world for good and for bad. The experiences of Auschwitz and Hiroshima compel us to disagree with the philosopher **Gottfried Wilhelm Leibniz** (1646-1716), who optimistically declared that our world is the best possible one. Even Augustine's Neoplatonic idea that evil is only a deficiency of the good no longer seems credible. Some would even agree with Richard Rubenstein when he ponders that if there were God, he could not have permitted Auschwitz to happen, and if he had, we would have to strip him of his divine office.[36] A few might even be inclined to agree with Arthur Schopenhauer when he rephrased Leibniz's assertion, saying that the world is "the worst of all pos-

35. William G. Pollard, *Chance and Providence* (New York: Scribner, 1958), 7.

36. Cf. Richard Rubenstein, *After Auschwitz: Radical Theology and Contemporary Judaism* (Indianapolis: Bobbs-Merrill, 1966), 87 and 153f. In the second edition he comes to a pantheistic understanding of God according to which God creates and destroys; cf. Richard Rubenstein, *After Auschwitz: History, Theology, and Contemporary Judaism*, 2nd ed. (Baltimore: Johns Hopkins University Press, 1992), 306.

184

sible worlds."[37] But it is not only God's special providence, or his miraculous activity, that is being challenged today; his general providence, or his benevolent presence with his creation, is also being questioned. How can God interact with the creative process, or accompany it in a benevolent way, if ultimately everything occurs "naturally"?

Surprisingly scientists today are raising again the question of teleology, assuming a directedness of nature, while theologians have largely abandoned the assertion that there is an empirically discernible teleology. Scientists, for instance, point out that the fundamental laws in physics would only have to be a little different from the way they are and life, as it is today, would never have been possible. It seems therefore that "a life-giving factor lies at the center of the whole machinery and design of the world."[38] Barrow and Tipler therefore introduced the weak anthropic principle in which they reasoned that taking into account the cosmological context, life could and indeed has evolved. Not everything in our world can be left to chance, but some things must be designed in such a way that life could evolve on the basis of carbon and that the universe is old enough that this development took place. Barrow and Tipler even ventured in the strong form of the anthropic principle the claim that the universe must have those properties which allow life to develop within it.

John Polkinghorne summarizes the main scientific insights assembled under the rubric of the anthropic principle as follows:

> Although life only began to appear on the cosmic scene when the universe was 11 billion years old, and self-conscious life when it was 15 billion years old, there is a real sense in which the cosmos was pregnant with life from the Big Bang onwards. The laws of nature were finely tuned in a way that alone made the evolution of carbon-based life a possibility. Only if the forces of nature were exactly what they are could there have been stars capable of burning reliably for the billions of years necessary to fuel the development of life on a planet. Only if the nuclear forces were exactly what they are would the first generation of stars have been able to make the chemical elements that are the

37. Arthur Schopenhauer, *The World as Will and Idea*, trans. R. B. Haldane and J. Kemp (London: Routledge & Kegan Paul, 1957), 3:395.

38. John A. Wheeler, in the foreword to *The Anthropic Cosmological Principle*, by John D. Barrow and Frank J. Tipler, 2nd ed. (Oxford: Clarendon Press, 1988), vii.

basis of life, so that in the death throes of a supernova explosion there spewed out the stardust of which we are made.[39]

Like all scientific explanations, the anthropic principle is directed to the past. Life in our universe has evolved and therefore, so the assertion, there must be certain constants that made the origin and development of life possible. In retrospect, the historical and evolutionary process is interpreted as a strictly immanent event. But what has actually happened and what took place at that point when it took place? The answer is disclosed neither by the anthropic principle nor by scientific research.

When we talk about divine providence, however, we do not turn to the past. Our attention is directed toward the present. The conviction is uttered that God has the present in his hand here and now and therefore our future is decided too. The assertion is made that the future will not open itself in any possible way. It will open only in the manner which is sanctioned by God. With this assertion we do not just focus on the cosmos but also on humanity and its conduct and history. Divine providence therefore extends to nature, to human conduct, and to history.[40] When we first turn to nature, we remember that all natural processes presuppose nature and matter. Yet these presuppositions cannot be taken for granted, because there is insufficient reason to suppose that an initial singularity occurred. The world is contingent, it is not absolute, and God is not in need of it. The world was created out of nothingness and is continuously threatened by it. Divine providence therefore asserts first of all that God continues to preserve his creation. Martin Luther, for instance, was much more impressed by the continuous preservation of God's creation than by the initial creative act. He remarked that many people start something but most do not have the energy to continue it.[41] God, however, maintains his creation and is with it in every single moment. Nothing is excepted from his caring providence.

39. John Polkinghorne, "Science and Theology in the Twenty-first Century," *Zygon* 35 (December 2000): 945.

40. These three areas of divine providence were also emphasized by Michael J. Langford, *Providence* (London: SCM Press, 1981).

41. Cf. Martin Luther, "Randbemerkungen Luthers zu den Sentenzen des Petrus Lombardus" (1510/11), in *Werke: Kritische Gesamtausgabe* (Weimar, 1883ff.), 9:66.29-34; hereafter cited as WA.

God maintains the oscillations of electrons as much as the encounter between two people that leads to marriage.

We are talking here about God who, as a Negro spiritual expresses, has "got the whole world in his hands." The world is enveloped by God who is present to it as a whole as well as to its individual entities. In perceiving God's relation with the world in this way, Arthur Peacocke particularly talked about a "top-down" causality in which God could be causatively effective without abrogating the laws and regularities that are operative in the world.

> Particular events could occur in the world and be what they are because God intends them to be so, without at any point any contravention of the laws of physics, biology, psychology, sociology, or whatever is the pertinent science for the level of description in question. . . .
>
> In thus speaking of God, it has not been possible to avoid talk of God "intending," of God's "freedom," that is, to avoid using the language of personal agency. For these ideas of "top-down" causation by God cannot be expounded without relating them to the concept of God as, in some sense, an agent, least misleadingly described as personal.[42]

God's preserving and life-furthering power is illustrated well by the Old Testament understanding of spirit *(ruah)*.[43] When God gives his spirit, something becomes alive and lives as long as God's spirit works in it. When God, however, takes away his spirit, then people or things perish. Therefore the psalmist pleads with Yahweh:

> Do not cast me away from your presence,
>   and do not take your holy spirit from me. (Ps. 51:11)

But if nothing can exist or occur without God, if God makes possible the earthquake as much as the life-giving spring rain, the destructive floods as much as a beautiful sunrise, do we then not equate divine providence with the laws of nature or with all natural occurrences? What

---

42. Arthur Peacocke, *Theology for a Scientific Age: Being and Becoming — Natural, Divine, and Human* (Minneapolis: Fortress, 1993), 159.

43. To the following, cf. Hans Schwarz, "Reflections on the Work of the Spirit outside the Church," *Neue Zeitschrift für Systematische Theologie und Religionsphilosophie* 23 (1981): 197ff.

sense would it make then to refer to God in our interpretation of nature if we pantheistically identify God with nature itself? Would not nature then become the highest principle without any mention of God?

Such questions would only be justified if we do not remember that God makes the difference between something and nothing. As Georg Wilhelm Friedrich Hegel (1770-1831), for instance, emphasized, God is not equal with nature but nature is dependent on God, it is contingent. If we do not understand God as the one who is behind *all* natural events, we run the risk of limiting God to ever fewer events for which we do not yet have a natural explanation. God then would become a cosmic jack-in-the-box. Yet John Polkinghorne cautioned that "the one god who is well and truly dead is the god of the gaps."[44] Much earlier Dietrich Bonhoeffer (1906-45) had warned us of the fate that such a God would encounter. He would be edged out of the world through our increasing knowledge of the natural causes and effects.[45] To avoid this danger Bonhoeffer emphasized that we must affirm God in the midst of life, in those places where we seemingly already know all the natural causes. This affirmation of God as the all-preserving power seems to be consequent and necessary for trusting in divine providence once we are confronted with the natural laws and the natural causes of all processes.

But if it is so, would we not affirm a demonic God, a God who plays dice with our future and who in dispassionate equanimity builds up and destroys? At this point the issue of theodicy or of God's justice emerges. In answering this question we must remember that the biblical witnesses do not divide historical and natural events into two categories: a category of evil for which Satan is responsible, and a category of good which is ascribed to God. They are much more interested in emphasizing that ultimately all processes serve to help God's kingdom triumph. This does not mean, however, that we are only informed about the journey itself, while the individual events up until the end are dark and enigmatic. But some events will indeed defy explanation. Altogether we can discern three areas with different parameters for explanations: nature, human conduct, and history.

44. Polkinghorne, "Science and Theology," 944.
45. Cf. Dietrich Bonhoeffer, *Letters and Papers from Prison,* ed. Eberhard Bethge, rev. and enlarged ed. (New York: Macmillan, 1968), 188, in a letter dated July 16, 1944.

Preservation within Nature

Nature serves as the foundation for all life and provides us with varying degrees of dependability which can be understood as the result of the caring activity of God:[46]

1. The first kind of dependability is represented in the rising and setting of the sun and in the cycles of the seasons. They provide the foundation for the development of life on earth and, as far as we know, are fully reliable. Of course, the movements of the planets, which can be calculated in advance, cannot be presupposed as absolutely reliable. Yet the probability for ultimate modes of behavior besides those we have observed over thousands of years is virtually zero. We remember that the reliability of the cycles of day and night and of the seasons is also reflected in the Noachic covenant. Yahweh assures us:

> As long as the earth endures,
>> seedtime and harvest, cold and heat,
> summer and winter, day and night,
>> shall not cease. (Gen. 8:22)

2. A different kind of reliability arises when several alternatives and large numbers of repeated incidences are involved. We encounter this, for example, in chemical reactions when wood or other fuels are burnt, in nuclear reactions in the core of the sun, and in the functioning of our body cells. Without the dependability of these processes, there would be no life on earth and no human history. Our very lives depend on the reliability of a huge number of these processes. The mutation of a single cell can lead to a dangerous cancer which damages our life severely or even cuts it short. Again the biblical witnesses remind us that God provides this kind of dependability. We hear that no sparrow falls to the ground without God's permission (Matt. 10:29), and all the hairs on our head are numbered (Matt. 10:30), and that God sends rain on the just and on the unjust (Matt. 5:45).

3. There is, however, a third kind of dependability which we encounter primarily in the evolutionary process of the universe and of life itself. "So far as the laws of nature and the structure of things in space and time are concerned, the universe *could* have had many histories than

46. To the following, cf. Pollard, *Chance and Providence*, 74-78.

the one it has had. At the same time, however, it is equally true, under the stern requirements of the necessity of choice in temporal existence, that it *can* have only one of these histories."[47]

Once the choice has occurred, all other possibilities are gone and lost forever. The whole evolutionary process, especially of life on earth, is so accidental that the American vertebrate paleontologist George G. Simpson (1902-84) could rightly claim that "there is no automatism that will carry him [humanity] upward without choice or effort and there is no trend solely in the right direction. Evolution has no purpose."[48] We remember that even Charles Darwin wished he could see a little more divine guidance in the evolutionary process. At the same time, Darwin advanced rather strict principles according to which the advancement of life and the evolution of new species proceeds. Scientific research has refined Darwin's findings considerably, but it has not repudiated them. Jacques Monod also said there is chance and necessity in the evolutionary process. Therefore this third kind of dependability tells us on the one hand that the natural process is trustworthy, since it is ongoing, but it tells us on the other hand that it is open and undetermined.[49]

The evolutionary process characterized by both chance and necessity also allows us to talk about God's continuing activity in the world.[50] In so doing, we dare not identify God with the physical processes lest we end up with a Spinozean equation of God and the world. God

47. Pollard, *Chance and Providence*, 68.

48. George G. Simpson, *The Meaning of Evolution: A Study of the History of Life and of Its Significance for Man* (New Haven: Yale University Press, 1960), 310. Of course, we do not want to say herewith — and neither does Simpson — that everything was completely unforeseeable in the evolutionary process. There are always covariances within this process and certain boundaries which cannot be passed or below which a certain species does not come (e.g., the minimum weight for mammals). This leads to an inner equilibrium within the evolutionary system with the goal of optimizing the conditions for existence and survival. Cf., for example, the paper of Paul Overhage, "Gebundene Mannigfaltigkeit," in *Gott in Welt: Festgabe für Karl Rahner,* ed. Herbert Vorgrimmler (Freiburg: Herder, 1964), esp. 842ff.

49. Simpson, 311, even states that this "new evolution involves knowledge, including the knowledge of good and evil." Humanity again is confronted with the basic choice between good and evil similar to the beginning in paradise.

50. For the issues relating to God's continuous creation, cf. the insightful analysis provided by Mark Worthing, *God, Creation, and Contemporary Physics* (Minneapolis: Fortress, 1996), 130-38 and 156f.

should also not be mistaken "as some kind of *additional* factor added on to the processes of the world."[51] Such a transcendent God would give the world too much independence. We should rather acknowledge that God works through the evolutionary process by being the provider of both continuity and openness. As John Polkinghorne appropriately asserts, God is present "in the chance as well as in the necessity of an evolving world."[52] God's continuous activity in the world does not just mean to determine what is to be. It also allows for freedom that creation can evolve in its own way. God is not a primordial tyrant at whose decree things come into being or proceed, but he is also a supporting and trusting God.

While the natural sciences have traditionally been seen as presenting a world that is predictable and governed by ironclad laws, we have noticed that especially in the twentieth century this perception has been severely shaken. "There are systems, at every level of complexity, whose future development is unpredictable. These were: subatomic systems, at the ('Heisenberg') micro-level; many-bodied Newtonian systems at the micro-level of description; and non-linear dynamical systems at the macroscopic level."[53] This new awareness of the unpredictability, open-endedness, and flexibility inherent in many natural processes and systems should not lead us to assume that this provides avenues for God interfering with natural causes. Instead it gives us a new appreciation for God's continuing creation. Not everything was predetermined from the very beginning. Creation as it presents itself to us today is indeed a miracle, not a necessity. The open-endedness also implies a certain precariousness. This in turn has challenged human imagination.

Some conclude from the openness of the evolutionary process that humanity should take this process into its own hands. By deciphering the genetic code we have more and more opportunities to do so. Yet humanity has always interacted with the evolutionary process. These interactions, however, do not change the overall impression of the process. Our own choice in the evolutionary process is part of the natural process and thereby is not exempt from accidental occurrences, for instance

51. Peacocke, 176.

52. John Polkinghorne, "The Life and Works of a Bottom-up Thinker," *Zygon* 35 (December 2000): 960.

53. Peacocke, 152.

in determining where we interact and how we do this. If chance would not have played a role in the great contributions for the future of humanities, many Nobel Prizes in the natural sciences, for peace, and in literature would not have been awarded to certain persons. We delude ourselves if we hope to be able to control the natural process in such a way that it becomes totally predictable. It is not our lack of insight that makes this undertaking impossible, but that the process of nature is "fundamentally uncontrollable."[54]

It is exactly with regard to this kind of dependability, this trustworthiness of the natural process, that we discern a peculiar train of thought in the biblical creation accounts. We hear that God initially said and it was so (Gen. 1:6f.). This is followed by the command for humanity to assume a certain responsibility (1:28). And finally we are told that God takes protective care of humanity (3:21: "And the LORD God made garments of skins for the man and his wife, and clothed them"). This means that the creative process and humanity's function within it do not suffice to make it dependable. Beyond initiating and maintaining his creation, God also has to abide with it. We are reminded here of Luther's observation that we should not think that God has retired and is sleeping on a pillow in heaven; instead he watches and guides everything. Luther was also much impressed with the fact that God did not abandon his creation after he made it. Luther wrote, "he [God] has not created the world like a carpenter builds a house and then leaves it and let it be the way it is, but he stays with it and sustains it the way he has made it, otherwise it would not remain."[55]

God's care for humanity as expressed in the natural process is not uniform. It proceeds on several layers of dependability involving greater or lesser freedom. Within this freedom adverse constellations have their place too, such as earthquakes and floods, or human management or mismanagement of the earth, and even to some extent the seductive and devastating antigodly powers. Of course, we could ask why God allows for freedom which does not preclude negative possibilities.

Yet what kind of freedom would that be that only contains good? If there were only freedom for good, that would only be coercion. There

---

54. So Pollard, *Chance and Providence,* 178.

55. Martin Luther, *Kaspar Crucigers Sommerpostille* (1544), in WA, 21:521.20-25, in a meditation on Rom. 11:33-36.

would be no actual freedom to choose between good and bad. God would be the primordial tyrant at whose fiat everything occurs. Since the overpowering providence expressed in Stoic or Islamic thought is foreign to the biblical experience, we arrive at a notion of providence that sets forth but does not compel, a notion of providence that accomplishes and does not dehumanize. The final goal and degree of dependability are known, and therefore the natural process is trustworthy though not foolproof.

In talking about the natural process we acknowledge that there is a dynamic drive in nature. Scientific investigation has shown us that nature moves consistently with and in space and time. Since, phenomenologically speaking, space and time are also nature's outermost parameters, we have no reference point to clearly discern the teleological direction of this movement. Yet when this drive is related to God, such a direction is provided. Now we recognize the eschatological dimension of nature and hear that "the creation waits with eager longing for the revealing of the children of God" (Rom. 8:19). We also realize that nature is not an isolated phenomenon. It is tied to and expressive of the phenomenon of life and therefore of humanity. Knowing about the God who cares for us through the natural process, we can approach the future confidently.

Preservation through Moral Conduct

Besides natural events, there is another important way in which God's general providence expresses itself. In the natural process God's continuous creative activity is dominant; however, in the moral process God's preserving creative activity rules supreme. God maintains and guides his creation, and within creation especially humanity, so as to avoid creation's destruction and self-annihilation. God did not abandon the world once he had created it. He continues to accompany it in a caring way. Similarly, he does not let humanity go on its own once it appears on the scene. God endows humanity with certain guidelines within which it can unfold itself and which may aid it in finding its proper place within creation. Yet how can we find these guidelines?

The most obvious place to seek guidance is one's conscience. The name "conscience" is derived from the Latin *conscire* and means "being a witness to" or "knowing with" someone. In other words, the conscience was originally understood as a kind of moral self-reflection that scrutinizes one's activities. When we recall, however, how many crimes have been committed in nationalistic and fascist countries in the name of conscience, we wonder what kind of normative and autonomous voice this conscience is. We realize that our conscience itself is not a moral norm; rather it attests to those norms it attempts to enforce.[56] Yet the extent to which these norms are perceived and enforced differs greatly. For some they seem to be hardly existent, while for others they are torturous, as they were, for instance, for the young Luther in his monastery cell. If we want to address ourselves to the trustworthiness of the moral process, we cannot just look at the conscience as the expression of moral norms. We must search behind the conscience for those very norms.

A promising way in our search for normative forces of human behavior opens up when we investigate the so-called natural law. The notion of a natural law goes back to the ancient idea that the gods gave people rules and regulations according to which people could and should conduct their lives. Aristotle cast this into a philosophical framework by claiming that basic to all human law is the divine law. The fundamental divine law is that which is just by nature and "which is the criterion and creative foundation of all human legislation and jurisdiction."[57]

At first the Christian community was rather hesitant to adopt the idea of a natural law, which was especially prevalent among the Stoics. It was primarily Origen (185-254) who paved the way for the reception of natural law in the Christian community when he identified Christ, the *logos*, with the rational, i.e., reasonable structure of the world. In Hellenism this rational structure was called *logos* also. At the same time, however, Origen pointed out that the existing positive law of the state could easily conflict with natural law. For instance, he claimed: "When the law of nature, that is, the law of God, commands what is opposed to the written law, observe whether reason will not tell us to bid a long fare-

56. Cf. the important collection of papers, Charles E. Curran, ed., *Absolutes in Moral Theology?* (Cleveland: Corpus Books, 1968), which contains Robert H. Springer's paper "Conscience, Behavioral Science and Absolutes," 19-56, in which he states that "greater relativity in the abstract will yield sounder moral conclusions in the concrete" (56).

57. So Emil Brunner, *Justice and the Social Order,* trans. M. Hottinger (New York: Harper, 1945), 6.

well to the written code, and to the desires of its legislators, and to give ourselves up to the legislator God, and to choose a life agreeable to His word, although in doing so, it may be necessary to encounter dangers, and countless labors, and even death and dishonor."[58]

Soon Origen and other theologians (such as Lactantius) argued, referring to the world reason, that to outlaw Christian congregations and to demand sacrifices to the gods was contrary to the natural divine law.[59] Thus for the first time an attempt was made to found and limit state laws through the divine natural law.

In Augustine we find a more explicit understanding of the natural law. He asserts now that we consider something just and right because nature teaches us. We do not consider it just and right because we have arrived at it through human convention. With this statement he almost anticipated the heavy critique of the natural law a thousand years later by English empiricists such as Thomas Hobbes (1588-1679).[60] "The eternal law is the divine order or will of God, which requires the preservation of natural order, and forbids the breach of it."[61] The eternal law is not temporal, but is concomitant with order, peace, and harmony. At the same time, this law is equal to the wisdom and will of God. The eternal law must be distinguished from the temporal law, regardless of whether it was given in paradise or implanted in our hearts or promulgated in writings.[62] The eternal law is eternally unchangeable and applies to the whole of creation.[63] Animals, however, are subjugated under this law in such a way that they have no part in it, while angels participate in it fully. Humanity stands in the middle, in part subjugated to it like animals and in part participating in it like angels. Natural law belongs essentially to our humanity. It is therefore made concrete and present in many ways, as for instance in the Golden Rule, the Mosaic Law, and the

---

58. Origen, *Against Celsus* 5.37, in the Ante-Nicene Fathers, 4:560.

59. Cf. Lactantius, *The Divine Institutes* 5.12, in Fathers of the Church, 49:356ff.

60. Cf. Thomas Hobbes, *Leviathan* (chap. 14), introduction by A. D. Lindsay (New York: E. P. Dutton, 1950), 107, who understands the natural law primarily from the perspective of a contract or of mutual agreement.

61. Augustine, *Reply to Faustus the Manichaean* 22.27, in Nicene and Post-Nicene Fathers, 1st ser., 4:283.

62. Cf. Augustine, *On the Psalms* 119.117, in Nicene and Post-Nicene Fathers, 1st ser., 8:580.

63. To the following, cf. Augustine, *Reply to Faustus* 22.28; 4:284.

law inscribed in our hearts. In these forms the temporal law is the re-flection of the eternal law.[64]

Thomas Aquinas (ca. 1224-74) largely followed Augustine in his understanding of natural law. For him "the Eternal Law is nothing other than the exemplar of divine wisdom as directing the motions and acts of everything."[65] Contrary to Augustine, Thomas does not think the term "law" *(lex)* is derived from reading *(legere)* or choosing *(eligere)*, but it "comes from *ligando*, because it is binding on how we should act."[66] He therefore emphasizes the normative element in the law. "The Law is nought else than an ordinance of reason for the common good made and promulgated by the authority who has care of the community."[67]

Thomas distinguishes four different kinds of laws: (1) divinely re-vealed laws, directed toward supernatural purposes; (2) positive hu-man laws based on natural law; (3) natural law itself; and (4) the world law from which natural law follows. Natural law, then, "is this sharing in the Eternal Law by intelligent creatures."[68] This means that natural law is connected with both eternal law and the natural judgment of human reason. Through human reason the eternal law gains binding power as a reasonable prescript. Natural law is therefore essentially reasonable.[69]

By basing natural law on eternal law *and* reason, Thomas not only ensured its binding character. Once reason took its own path, as hap-pened during the Enlightenment, many things were introduced as rea-sonable which were clearly against any divine (eternal) law. We notice this to some extent in Thomas himself when, following Aristotle, he ar-gues for slavery as something given by nature. Slavery follows from the fact that "it is expedient for him [the slave] to be ruled by a wiser whom he serves."[70] Thereby Thomas sanctions the feudal society in existence at his time. Though he could have designed a very different order of so-

---

64. Cf. Augustine, *Letters* 157, in Fathers of the Church, 20:331.
65. Thomas Aquinas, *Summa Theologiae* 1a2ae.93.1, Blackfriar ed., 28:53.
66. Thomas Aquinas, *Summa Theologiae* 1a2ae.90.1; 28:7.
67. Thomas Aquinas, *Summa Theologiae* 1a2ae.90.4; 28:17.
68. Thomas Aquinas, *Summa Theologiae* 1a2ae.91.2; 28:23.
69. Thomas Aquinas, *Summa Theologiae* 3a2ae.93.5; 28:65ff.
70. Thomas Aquinas, *Summa Theologiae* 2a2ae.57.3; 37:13. Cf. also the comprehen-sive study by Bénézet Bujo, *Moralautonomie und Normenfindung bei Thomas von Aquin. Unter Einbeziehung der neutestamentlichen Kommentare* (Paderborn: Schöningh, 1979), 293ff.

ciety, Thomas has not yet developed a universally applicable natural law for human freedom. Since natural law results primarily in reasonable action according to body, soul, and mind, humanity sets up natural laws on the basis of the drive to self-preservation. It establishes natural laws according to its animal nature, which it shares with all other sentient beings, for the purpose of procreation and nurture of its descendants. According to its reasonable nature, humanity has a natural inclination to recognize God and to live in community. Once it has been determined what the essence of humanity is, the law must be developed accordingly to further our humanity. Of course, besides the natural law of the individual, Thomas does not neglect the natural law of the community as it applies to family, state, and church.

When Martin Luther says in his *Lectures on Romans* that "we make up many stories about the Law of nature," we might conclude that he did not much appreciate this time-honored concept.[71] When we also remember that he called reason a whore, this might reinforce our picture of Luther's low esteem of humans as reasonable beings.[72] But Luther did not reject reason and natural law. He simply wanted to point to the problems contained in both.

Luther was convinced that God had ordered everything in the world, from the largest events to the smallest details. In his *Lectures on Galatians* of 1531, Luther stated with reference to Deuteronomy 22:5: "The male was not created for spinning; the woman was not created for warfare."[73] God has established a certain order in creation according to which each member has its specific place and function. Let the king rule, the bishop teach, and the people obey the magistrate. "In this way let every creature serve in its own order and place."[74] This order extends over the whole of creation from nature to humanity itself. If God has ordered nature, then it is only logical that even interhuman relationships are not arbitrary. Indeed, Luther is pointing out framework-type structures for family, government, and other interhuman relation-

---

71. Martin Luther, *Lectures on Romans: Glosses and Scholia* (1515/16), in *LW*, 25:344, in his interpretation of Rom. 8:3.

72. Cf. Martin Luther, *Predigten des Jahres 1546*, in *WA*, 51:126.9f., where Luther says reason is "the biggest whore that the devil has."

73. Martin Luther, *Lectures on Galatians* (1535), in *LW*, 26:307, in his explanation of Gal. 3:19.

74. Luther, *Lectures on Galatians*, 26:308.

ships. For instance, God has instilled in the hearts of parents that they serve their children to the best of their ability, nourish them, care for them, and bring them up with great diligence.[75]

Luther claims that these orders of relationships between people are best expressed in natural law. For instance, everybody knows that we should obey our parents, since we are from the same blood and since they bring us up.[76] Likewise reason tells us that we should not kill anybody. From passages such as Matthew 7:12 and Romans 2:15 Luther concludes that the natural law contains all the precepts of the prophets and all other commandments.[77] To a large extent Luther thinks of natural law as identical with the Mosaic Law. Since the Mosaic Law, however, was given to the Jewish people, only that part of this law which is congruous with the natural law is obligatory for everyone. The matter is different with the commandment of love. Luther rightly recognizes that it is part of the natural law.

Since the "natural law is inborn in us like the heat in fire and the fire in flint, . . . it cannot be separated from divine law."[78] It is a natural gift of God which was not lost through the fall. It enables and facilitates communal life and serves as the basis and corrective for all other forms of law. Luther states that if natural law and reason were inherent and available to all, all people would be equal in their conduct.[79] However, he concedes that natural law is no longer available in an absolutely normative form. While it is still inscribed into our hearts, because of our sinfulness its words do not have compelling power.[80]

Luther similarly assesses the power of reason. For instance, he admonishes a prince to "determine in his own mind when and where the law is to be applied strictly or with moderation, so that law may prevail

---

75. Martin Luther, *A Simple Way to Pray* (1535), in *LW*, 43:203, in his explanation of the fourth commandment.

76. Martin Luther, *Predigten über das 2. Buch Mose* (1524-27), in WA, 16:512.3-6, in his explanation of the third commandment.

77. To the following Martin Luther, *Against the Heavenly Prophets, in the Matter of Images and Sacraments* (1525), in *LW*, 40:97f.

78. Martin Luther, "Table Talk" (2119 B; 1531), in WA *Tischreden* 2:374, 17ff.

79. Martin Luther, *Psalm 101* (1534/35), in *LW*, 13:161, in his explanation of Ps. 101:1.

80. Martin Luther, *Lectures on Deuteronomy* (1525), in *LW*, 9:108, in his exegesis of Deut. 10.

at all times and in all cases, and reason may be the highest law and the master of all administration of law."[81] But reason does not function independently as the norm for natural law, though Luther admits that the written law stems from reason as the wellspring of all law.[82] Natural law can fulfill its normative function only insofar as it is embedded in the ordering activity of God which surrounds and maintains it and which reason continuously presupposes.[83] Reason does not inaugurate the trustworthy moral process, but it discovers and forms the legal process according to God's ordering action. As long as reason and natural law are sustained by the ordering activity of God, they are the normative and formative forces of the moral process.

In the Enlightenment, however, Luther's admonition concerning the relationship between God's ordering activity and the natural law was discarded. The notion developed that natural law could be maintained on a strictly rational basis without reference to God. For instance, Jean-Jacques Rousseau (1712-78) claimed that humanity is by nature good. Furthermore, humanity could regain this natural state if it would turn away from the corrupting influences of society and civilization and be itself, instead of simply being citizens whose value depends on the community in which they live.[84] This idea is pursued further in his *Social Contract* (1762). While he realized that human society is threatened by the tyranny of the people themselves, he was convinced that a true and just society cannot be based on mere force; it must rather be consistent with people as free and rational beings. Therefore he suggested that humanity be liberated from the tyranny of the individual human will. Rousseau insisted that "each one of us puts into the community his person and all his powers under the supreme direction of the general will; and as a body, we incorporate every member as an indivisi-

---

81. Martin Luther, *On Temporal Authority: To What Extent It Should Be Obeyed* (1523), in *LW*, 45:119.

82. Luther, *On Temporal Authority*, 45:129.

83. Cf. Luther's argument in *The Judgment of Martin Luther on Monastic Vows* (1521), in *LW*, 44:336.

84. Jean Jacques Rousseau, *Emile or Education*, trans. Barbara Foxley (New York: E. P. Dutton, 1948), 5ff.; cf. also William Boyd, *The Educational Theory of Jean Jacques Rousseau* (New York: Russell & Russell, 1963), 190, who rightly claims that in *Emile* Rousseau does not neglect nature. "He is merely seeking for a method of keeping men as near nature as possible under existing social conditions."

ble part of the whole."[85] Freed from the narrow confines of its own being, a human being will find fulfillment in a truly social experience of fraternity and equality with citizens who share the same ideals.

The question which must be posed is whether one can count on such agreement by nature. Is the foundation for Rousseau's "social contract" really to be found in nature? In his *Essay concerning Human Understanding* (1690), John Locke (1632-1704) had already asserted that all our knowledge only stems from sense perception.[86] Following this empirical trend, David Hume (1711-76) finally discarded all metaphysics. Though he still allowed for an original moral sense in humanity, he distinguished between matters of fact, which rest on certain conventions of experience, and truths of reason, which are based on certain conventions of reason. There is no longer an eternal norm of justice. Law becomes a matter of convention.[87] Yet if there is no longer a directing moral influence, how can sinfully inclined human beings, whom Rousseau portrays, become citizens who strive for fraternity and equality? Humanity seems to be on a course leading to enslavement rather than freedom.

When ontologically grounded natural law was abandoned, not only its normative character but also its outdated manifestations were discarded. For instance, the idea was increasingly challenged that there were different classes of people with different rights and privileges. Undeniable progress could now be made by instituting inalienable human rights for all people and democratic procedures for dealing with each other. Yet in doing away with the metaphysically grounded natural law, one was also prone to rejecting the notion that natural law is more than the expression of circumstances or mere convention. As the Reformed theologian Emil Brunner (1889-1966) has shown, if there is nothing which is universally valid and no justice beyond ourselves that meets all of us as an undeniable demand, then there is no actual justice, "but only

85. Jean Jacques Rousseau, *The Social Contract* (1.6), translated, edited, and introduction by Maurice Cranston (Baltimore: Penguin Books, 1969), 61.

86. Cf. John Locke, *An Essay concerning Human Understanding* (2.1.2), ed. Alexander C. Fraser (Oxford: Clarendon Press, 1894), 1:64, after having rejected the concept of innate ideas.

87. For Hume, cf. Hans Schwarz, *The Search for God* (Minneapolis: Augsburg, 1975), 76ff.; and for the empirical foundation of laws, cf. David Hume, *A Treatise of Human Nature* (3.3.1), ed. L. A. Selby-Bigge (Oxford: Clarendon Press, 1896), 591.

power organized in one fashion or another and setting itself up as law."[88] The alarming frequency with which totalitarian systems have emerged during the last one hundred years should make us wonder whether reason itself is a sufficient foundation for the moral process. Even during the height of absolutism, kings and princes understood that they enjoyed their rule through the grace of God. But today's dictators exercise their rule in their own name. They no longer feel subject to higher powers. The individual human being has become its own ultimate measure for right and wrong.

But we should also remember Charles Darwin's assertion that human moral and mental faculties differ in degree rather than in kind from the capacities of animals. If there is indeed a moral continuity between animals and humans, there should be even more reason to assume that such continuity exists between one human being and another. Perhaps the reason why we tend to deny the basic unity of the human moral process lies in our sinful estrangement from each other. We are so far apart from each other that we tend to forget our common history and common destiny.

Wolfgang Wickler (b. 1931), a former student of Konrad Lorenz (1903-89), points out our common moral history in his interesting book *The Biology of the Ten Commandments*. He claims that the Ten Commandments (specifically the fourth to the tenth) are demands that have not just emerged on the human level. For instance, inhibition against killing of members of one's own group, against theft and lying, and the summons to honor older members of the same social group are already present among animals.[89] This does not mean that these commandments are immutable. As a foundation and summary of appropriate moral behavior, they show a development even within the human family.

There are Decalogue-like collections of commandments known in Egypt, India, and even among the Masai in Kenya.[90] Especially the lat-

---

88. Brunner, 8.

89. Cf. Wolfgang Wickler, *The Biology of the Ten Commandments,* trans. D. Smith (New York: McGraw-Hill, 1972), 76, 12, and 160f.

90. Cf. Siegfried Morenz, *Egyptian Religion,* trans. A. E. Keep (London: Methuen, 1973), esp. 112, where he refers to the so-called Book of the Dead, and 121ff., where he classifies the commandments in ancient Egypt as wisdom literature. This comment is especially telling, since it attests to the acquisition of right conduct through insight and not just divine revelation. Cf. also H. Saddhatissa, *Buddhist Ethics: Essence of Buddhism*

ter are instructive, since they portray a nomadic existence which necessitated significant changes in the Decalogue to adjust to the peculiarities of this form of life.[91] We are reminded here of Luther's insight: "What God has given from heaven to the Jews through Moses, he has also inscribed in the hearts of all humanity."[92] Luther recognized the Decalogue as a basic norm which is known to everyone even without the Mosaic Law. Ethnological research has adduced more and more evidence that indeed there are some basic norms of human behavior. The biologist Arnold W. Ravin from the University of Chicago expressed this very appropriately when he said:

> Every culture has a concept of murder, that is, a specification of conditions under which homicide is unjustifiable. Every culture has a taboo upon incest and usually other regulations upon sexual behavior. Similarly, all cultures hold untruth to be abhorrent, at least under most conditions. Finally, all have a notion of reciprocal obligation between parents and their children. These universal or near-universal ethics . . . do indicate some profound and fundamental needs in all men to behave within certain limits or ethical boundaries.[93]

This means that by nature the behavior of human beings is not as free and unspecified as we might initially assume. To be a human being means to act according to certain norms that enable us to live together and further our own species. For Arnold Ravin as well as for Konrad Lorenz and his followers, morality is natural. Sociobiology, however, cautioned against such optimism and showed that altruism among members of the same species is the result of the fact that the actual carriers of biological evolution are not individuals, whether species or single members, but the genes that cooperate in order to survive. Edward O. Wilson (b. 1929) therefore writes: "The genes hold culture on a leash. The leash

---

(London: George Allen & Unwin, 1970), 87; and Kashi Nath Upadhyaya, *Early Buddhism and the Bhagavadgita* (Delhi: Motilal Banarsidass, 1971), 413f. Cf. also Moritz Merker, *Die Masai* (New York: Johnson Reprints, 1968 [1910]), 335f. Of course, the Masai are a Semitic tribe, which Merker emphasizes, pointing to their common heritage with the Israelites.

91. Cf. Wickler, 44f., who refers here to Merker's study of the Masai.

92. Luther, *Predigten über das 2. Buch Mose,* WA, 16:380,19f., in a sermon on Exod. 19.

93. Arnold W. Ravin, "Science, Values, and Human Evolution," *Zygon* 11 (June 1976): 151.

is very long, but inevitably values will be constrained in accordance with the effects on the human gene pool."[94] The findings of behavioral genetics are not, however, totally void of moral implications.[95] It is still concerned with survival. Yet in contrast to Darwin, who considered the survival of individuals and of groups, behavioral geneticists have realized that individuals certainly do not survive, and groups quite often do not survive either. In an evolutionary scheme, only genetic units have an inherent tendency to last long enough to survive, and these units have evolved to survive by helping their copies reproduce wherever they may live. In order to succeed with propagation, certain behavioral traits are favored while others are abandoned as being unsuccessful. For humanity this would mean that the explicit forms of behavior depend upon the environment in which the social behavior takes place.

The trustworthiness of the moral process which is maintained by living according to these norms depends on our ability to develop these norms in such a way that they continue to be normative for interactions with the environment in which we live. If these norms are not developed in accordance with the changes we encounter and inaugurate, we will become helpless victims of the developments which surround us.[96] For instance, if we do not adjust our conduct to the crowded conditions in our modern anonymous mass society, this will lead to a deterioration of individual partner relationships.[97] We will "get on each other's nerves" to such an extent that either society completely disintegrates or new individual protective areas are created. Contrary to our obligation of governing the world and using it in accordance with the moral norms appropriate to us and our situation, the world would subdue us and impose on us a moral behavior that is appropriate to us and our environment. Instead of being agents of the conditions we encounter, we could at best react to them.

These insights and conclusions have important implications for our understanding of the traditional theology of the orders of creation. The Lutheran theologian Werner Elert (1885-1954) is certainly right when

94. Edward O. Wilson, *On Human Nature* (Cambridge: Harvard University Press, 1978), 167.

95. Cf. Hans Schwarz, "The Interplay between Science and Theology in Uncovering the Matrix of Human Morality," *Zygon* 28 (March 1993): 69f.

96. Cf. Wickler, 171ff.

97. Cf. Wickler, 179.

he claims that the orders of creation presupposed in the Decalogue with words of command and prohibition are orders of God's creation to which we belong as created beings. Therefore he calls them orders of creation.[98] We wonder, however, whether Elert is correct when he asserts that these orders are always orders that already exist and not orders that should be maintained. Truly, the sixth commandment does not constitute marriage but presupposes it. Adultery was committed through acts that were considerably different in King David's Israel from those in Martin Luther's Germany or in Puritan England. Needless to say, in each epoch marriage was understood quite differently. This does not mean that the underlying norm "thou shalt not commit adultery" had no binding value. But if we want to maintain the goal of a durable relationship which is envisioned in this commandment, then the exact interpretation of its normative ought-character must be reconsidered with each changing situation.

Since moral norms are goal oriented, intended to preserve rather than constitute the species, it might be good to follow the suggestion of Walter Künneth (1901-97). He speaks of orders of preservation instead of orders of creation. Künneth does not intend to diminish the creational character of these orders, but he objects against their static interpretation and maintains that creation must be perceived from the perspective of God's conserving activity.[99] This is even more necessary since we know God's creation only in its fallen condition, under the aspect of preservation. According to Künneth, the orders of preservation counteract the tendencies of the destructive antigodly powers. They are a sign that God does not want to destroy the world but conserves it for Christ's sake and toward Christ. The orders of preservation therefore ultimately have eschatological character: they point to the eschatological fulfillment in the new creation.

Since these orders are no longer evident in their original divine creational intention, they assume the character of a law. But as law they enable and facilitate the living together of people. They express a mutual obligation and therefore have basically the character of mutual ser-

---

98. Cf. Werner Elert, *The Christian Ethos,* trans. Carl J. Schindler (Philadelphia: Muhlenberg, 1957), 77ff.

99. For the following, cf. Walter Künneth, *Politik zwischen Dämon und Gott: Eine christliche Ethik des Politischen* (Berlin: Lutherisches Verlagshaus, 1954), 139f.

vice. Yet they are also susceptible to sinful distortion and can be misused to perpetuate injustice and inequality. Therefore they should never be separated from God as the originator and granter of these orders. This does not mean that we should go as far as Karl Barth, who claims that through its sinful existence humanity is in such depravity that it cannot, by its own power, know anything about these fundamental moral laws of nature. Barth therefore insists that this order cannot be found anywhere. It has "sought us out in the grace of God in Jesus Christ revealed in His Word, disclosing itself to us as such where we for our part could neither perceive nor find it."[100]

Bonhoeffer attempts a similar christomonistic foundation of these orders as does Barth. Bonhoeffer claims it is an empty abstraction to talk about the world if one does not relate it to Christ.[101] The relationship "of the world to Christ assumes concrete form in certain mandates of God in the world. The Scriptures name four such mandates: labour, marriage, government and the Church." The term "mandates" is very apt, for it points to the obligatory aspect of these orders. Yet we are afraid that an exclusively christocentric approach to these orders — whether we accept their foundation in Christ or not — obscures the fact that the acknowledgment of certain moral norms is a necessary condition for human existence as such.[102] Since they are binding for both atheists and Christians, they are not a source of revelation.

Atheists can recognize these norms of moral behavior as partly inborn and partly handed on by tradition. In the light of God's self-disclosure in Jesus Christ, however, Christians perceive in these norms the preserving activity of God. They realize that through the evolving moral process God is preserving the human community against human self-destruction and against the destructive and seductive tendencies of the antigodly powers. Luther expressed this very picturesquely when he

100. Karl Barth, *Church Dogmatics,* vol. 3/IV: *The Doctrine of Creation* (Edinburgh: T. & T. Clark, 1958), 45.

101. Cf. Dietrich Bonhoeffer, *Ethics,* ed. Eberhard Bethge, trans. N. H. Smith (New York: Macmillan, 1965), 207, including the quote.

102. Immanuel Kant in his *Metaphysical Foundations of Morals,* in *The Philosophy of Kant: Immanuel Kant's Moral and Political Writings,* edited with introduction by Carl J. Friedrich (New York: Random House, 1949), 150f., argues on the basis of "pure reason" that there are certain moral norms which, when universalized, will support human existence while others, once universalized, will impede it.

said: "If God were to withdraw his protective hand and leave room to the devil you would become blind, or an adulterer and murderer like David. You would fall and break your leg and drown."[103] Through God's self-disclosure in Jesus Christ we are also reminded of the trustworthiness yet transitory character of the moral process. It will find its fulfillment and completion in the eschatological new creation, when the moral norms will be unimpaired and self-evident. Then our moral behavior will be characterized by complete harmony with God.

It has not been difficult for us to assert the trustworthiness of the natural process. We know that the universe has evolved in an amazing way over the course of its existence, and there is little doubt that this process of unfolding will continue. Yet amid all the present-day uncertainties it has been more difficult for us to assert that there are basic moral norms which we can trust and which we must continuously develop. When we consider the historical process, then its trustworthiness is even more difficult to maintain. We are informed by the Christian tradition that God is the Lord of history, but when talking about human history the question immediately emerges: Whose side is God on?

Preservation through the Historical Process

In World War I the churches called upon God to bless the weapons on both sides of the trenches. After the Allies had dictated to Germany the terms of peace in the Treaty of Versailles (1919), the "German Christians" saw the special finger of God in Hitler's rise to power in 1933. Even in 1934 when many had discovered that Hitler was not a blessing, a group of theologians in Württemberg (Germany) wrote the following statement: "We are full of thanks to God that he, as Lord of history, has given our country Adolf Hitler, as leader and savior from our difficult lot. We acknowledge that we are bound and dedicated with body and soul to the German state and to its *Führer*. This bondage and this duty contain for us, as evangelical Christians, their deepest and most holy demand in the fact that it is obedience to the commandment of God."[104]

---

103. Martin Luther, *Predigten des Jahres 1531,* in WA, 34 II:237.16-19, in a sermon on the festival of St. Michael's.

104. "Zwölf Thesen der Kirchlichen Einheitsfront in Württemberg" (May 11,

Even as a theologian who was never associated with the so-called German Christians, Gerhard Kittel (1888-1948), the former editor of the *Theological Dictionary of the New Testament,* confessed that he had prayed for years that his people might be saved from their distress and disgrace.[105] Should the emergence of Hitler not be regarded as God's answer to such prayers? Had not Hitler himself invoked the power of "Providence" when he was amazingly spared from assassination attempts, claiming "that he was a man chosen by Providence to act as the agent of the 'World Historical Process'"?[106]

But it was not only Christians in Nazi Germany who ventured such "providential" interpretations of history. In 1941, when the invasion of the German army in Russia was most threatening, Patriarch Sergius, the metropolitan of the Russian Orthodox Church in Moscow and Kolomna, confidently wrote in a circular addressed to the laity and clergy of his district that, as on previous occasions, "with the help of God, they [i.e., our people] will once again chase away the troops of the enemy."[107] Of course, we could argue with Luther that God simply "uses one rascal to punish the other."[108] That is, God does not condone either of them but simply uses them in his providential activity to restore order and justice. The factor of historical survival or eminence does not legitimate a certain power or authority as providential.

There are still amazing historical events, however, that invite true theological evaluation. For instance, when the Gothic hordes of King Alaric (ca. 370-410) ransacked Rome in 410, many pagans claimed that the ransacking signaled the wrath of the ancient gods.[109] These gods

---

1934), in *Die Bekenntnisse und grundsätzliche Äußerungen zur Kirchenfrage,* vol. 2, *Das Jahr 1934,* ed. Kurt Dietrich Schmidt (Göttingen: Vandenhoeck & Ruprecht, 1935), 73.

105. Cf. Gerhard Kittel, in Karl Barth and Gerhard Kittel, *Ein theologischer Briefwechsel* (Stuttgart: W. Kohlhammer, 1934), 12.

106. Alan Bullock, *Hitler: A Study in Tyranny,* rev. ed. (New York: Harper Torchbook, 1964), 723. Bullock comments: "Anything, however trivial, which went right in the last two years of the war served Hitler as a further evidence that he had only to trust in Providence and all would be well."

107. Sergius, *Die Wahrheit über die Religion in Russland,* trans. Laure Wyss (Zollikon-Zürich: Evangelischer Verlag, 1944), 16.

108. Cf. Martin Luther, *Admonition to Peace: A Reply to the Twelve Articles of the Peasants in Swabia* (1525), in *LW,* 46:32.

109. Cf. Etienne Gilson, introduction to *The City of God,* by Augustine, ed. Vernon J. Bourke (Garden City, N.Y.: Doubleday, Image Books, 1958), 16.

were angry because the people had forsaken the ancient cults and adopted Christianity. However, the Church Father Augustine (354-430) attempted to show in his book *The City of God* that this was not the case. He claimed that the decline and fall of Rome were due to the moral depravity of paganism. A generation later the presbyter Salvian of Marseilles explained the decline of Rome as a judgment not on the heathen but on the Christians.[110] The fall of Rome was God's punishment for the unrighteousness of the Christians in church and society. What should we then trust, the interpretation of the pagans, or of Augustine, or of Salvian? Perhaps we should first realize "that the interpretation of an historical event as a special revelation of Providence too easily becomes a piously disguised form of self-justification."[111] It is one thing to believe that through God's providential action, even in the historical process, his kingdom will finally triumph, but it is yet another to conceive divine providence atomistically as a fragmentary demonstration of his power.

We are unable to claim with Friedrich Schiller (1759-1805) that "world history is world judgment."[112] It is also difficult to agree with the attempt of G. F. W. Hegel to conceive of world history as a theodicy of God in which the thinking spirit is "reconciled with the fact of the existence of evil."[113] Yet should we follow Dorothee Sölle (b. 1929) when she claims that the pain, injustice, and suffering of the innocent lead to the dethronement of God almighty, the king, father, and ruler of the whole world?[114] Has not Dietrich Bonhoeffer claimed that "God lets himself to be pushed out of the world" so that "God is weak and

110. Salvian, *The Governance of God* 6.12 and 7.1, in Fathers of the Church, 3:172 and 185.

111. G. C. Berkouwer, *The Providence of God,* trans. L. Smedes (Grand Rapids: Eerdmans, 1952), 180.

112. See Friedrich Schiller's poem "Resignation" (1786), in *Gesammelte Werke in fünf Bänden,* ed. Reinhold Netolitzky (Gütersloh: C. Bertelsmann, 1959), 3:394.

113. Georg Friedrich Wilhelm Hegel, *The Philosophy of History,* preface by Charles Hegel, trans. J. Sibree, introduction by C. J. Friedrich (New York: Dover, 1956), 15. It is also difficult to agree with him when he writes: "God governs the world; the actual working of his government — the carrying out of his plan — is the History of the World" (36).

114. Cf. Dorothee Sölle, *Christ the Representative: An Essay in Theology after the "Death of God,"* trans. D. Lewis (Philadelphia: Fortress, 1967), 150f., where she quotes Bonhoeffer's statement in our following quotation.

powerless in the world"?[115] Was William Hamilton (b. 1924) right when he claimed that he carried out the legacy of Bonhoeffer in asserting that the traditional sovereign and omnipotent God is difficult to perceive or meet? Hamilton suggested that "in place of this God, the impotent God, suffering with men, seems to be emerging."[116] Finally Thomas J. J. Altizer (b. 1927) speaks of the death of God as an event in history. "We must realize," he states, "that the death of God is an historical event, that God has died in our cosmos, in our history, in our *Existenz*."[117]

It would indeed be futile to attempt a justification of the historical process as a result of God's providential activity. Luther's emphasis on the theology of the cross should serve as a warning against such endeavors. In the Heidelberg Disputation (1518) he asserts: "True theology and recognition of God are in the crucified Christ."[118] Luther realized that God completes when he destroys, "that he makes alive when he puts one on the cross, that he saves when he judges," that he discloses himself when he disguises himself.[119] Luther consequently concluded that God works under the appearance of the opposite. Bonhoeffer emphasized this "weakness" of God too. But unlike some of his followers, he also confessed in his *Letters and Papers from Prison*, very much like Luther: "I believe that God is no timeless fate, but that he waits for and answers sincere prayers and responsible actions."[120]

From what has been written so far, we should not conclude that acknowledging God's providential involvement in the historical process is simply a matter of faith. If this were so, it could easily lead to credulity and to an attitude of regarding any historical oddity as the result of God's will. This could be even more true in history than in nature. In several respects the causal explanations of historical events

115. Bonhoeffer, *Letters and Papers*, 188, in a letter of July 16, 1944.

116. William Hamilton, *The New Essence of Christianity* (New York: Association, 1966), 54, where he refers to Bonhoeffer's quotation mentioned in the above footnote.

117. "America and the Future of Theology," in Thomas J. J. Altizer and William Hamilton, *Radical Theology and the Death of God* (Indianapolis: Bobbs-Merrill, 1966), 11.

118. Cf. Martin Luther's explanation to Thesis 20 of the Heidelberg Disputation of 1517, in *LW*, 31:53.

119. So Erich Seeberg, *Grundzüge der Theologie Luthers*, 2nd ed. (Stuttgart: W. Kohlhammer, 1950), 54f.

120. Bonhoeffer, *Letters and Papers*, 11, from a 1943 essay.

contain a significantly higher degree of subjectivity than those in the natural sciences:[121]

1. In the natural sciences an established causal connection between two stages of experience can be repeated by anyone at any time. Historical events, however, are singular, and their presumed sequence cannot be subjected to experimental scrutiny (e.g.: Does World War I always lead to World War II?).

2. In the natural sciences paradoxical phenomena (e.g., duality of wave and corpuscle) can be explained as complementary. Differing historical perspectives, however, usually exclude each other (e.g., the attempts to explain the reason for the devastation of Rome in 410).

3. In the natural sciences the causal sequence between two stages of experience can always be established. If hypothetical forces have to be assumed to establish such a sequence, such procedure usually leads to new advancements in science. With historical events the causal sequence is often not evident and history appears as a sequence of accidents (e.g., the legal phrase "an act of God").

Since we can justify God's providence neither by pointing to historical accidents nor by simply equating in deterministic fashion the historical process with God's providence, a criterion to assess the trustworthiness of God's providential activity in the historical process must be gained from other sources.

It is important here to remember that God is introduced in Scripture as the one for whom nothing is impossible and who determines the course of history. This confession of God's almighty power and his predetermination of history, however, only serves to emphasize the conviction that God will bring his plan of salvation to completion (Gen. 18:14 and Luke 1:37). This confession does not impress the notion upon us that all of history runs according to God's preconceived plan. When Jesus was asked for a theological interpretation of the death of the Galileans that Pilate had slain, or of the eighteen men who were killed by the tower that collapsed at Siloam (Luke 13:1-5), he did not equate the course of history with God's judgment. Jesus reacted similarly when his disciples asked him why a man was born blind (John 9:3). At the same time, Jesus did not simply shrug his shoulders and declare that

---

121. For the following, cf. Albert C. Outler, *Who Trusts in God: Musings on the Meaning of Providence* (New York: Oxford University Press, 1968), 45f.

history is without meaning. In the first two instances he answered in existential fashion that events like these remind us of our own mortality and sinfulness. In the latter case he commented, according to the Evangelist, that this case serves to make the works of God manifest.[122]

In other words, historical events are not an end in themselves. They are also not just part of the larger context of world history. Ultimately historical events have eschatological significance. They are "a living reminder of the End, speaking sometimes with certainty and more often in utter ambiguity."[123] Though the New Testament alerts us to watch for the signs of the end, it is obvious to the New Testament writers that history does not provide us with a timetable for events leading to the eschatological fulfillment of history and creation (Matt. 24:32f. and 36f.).

While we must agree with the biblical testimony that the historical process will find its fulfillment and completion in the eschaton, we must refrain from the temptation to identify God with one of the causes of the historical process or with the cause. Friedrich Gogarten (1887-1967) is more convincing than such simplistic solutions. He reminds us that God has accepted us in his sonship. In so doing God has granted us the freedom to lead our lives responsibly without interfering in our lives through divine providence.[124] God wants us to be his governors in the world. A governor without freedom and responsibility would be a mere puppet and not a responsive and responsible being. Yet responsibility does not exclude divine providence. On the contrary, it necessitates providence. We should be God's governors, and because of this responsible God-relationship we need God's guidance. This can be illustrated by the Old Testament understanding of *berith* (covenant) which also plays an essential role in the Christian faith. We see how important the notion of covenant is when we remember that "testament" as used in Old or New Testament really means covenant. In Old Testament thinking a covenant "mostly involves a promise by the master or suzerain to take his vassal under protection," but it is not an "agreement or settle-

122. Cf. Rudolf Bultmann, *The Gospel of John,* trans. G. R. Beasley-Murray (Philadelphia: Westminster, 1971), 331, esp. n. 3, in his exegesis of John 9:1-7.

123. Evgenii Lampert, *The Apocalypse of History: Problems of Providence and Human Destiny* (London: Faber and Faber, 1948), 176, who convincingly asserts an apocalyptic and eschatological interpretation of history.

124. Cf. Friedrich Gogarten, *The Reality of Faith: The Problem of Subjectivism in Theology,* trans. Carl Michalson (Philadelphia: Westminster, 1959), 55ff.

ment between two parties" or "a mutual agreement."[125] Finite humanity attempting to fulfill its task as God's governor is this inferior agent who needs God's assistance and guiding providence.

We have seen that God grants his assistance and guiding providence in providing the trustworthiness of both the natural and the moral process. He also grants the trustworthiness of the historical process. However, human life and history in any form are unwarranted. The believer does not perceive it as an accident. Human life and history are the result of God's creative and sustaining activity. The believer receives this gift unmerited and undeservedly. Yet the believer does not understand this trustworthiness (of the historical process) to result from the God who provides order. Certainly the impact of the natural and moral processes on history maintains order and averts chaos. But the primary thrust of the historical process is directed toward grace and fulfillment, and not toward order and sustenance. The unresolved tensions of history illustrate this especially well. God reminds us that we "groan inwardly while we wait for adoption" and that the whole of creation still waits "with eager longing for the revealing of the children of God" (Rom. 8:19-23). We recognize that our world is in transition from "in the beginning God created heaven and earth" to "and God will be all in all." Consequently our position as God's administrators of this world and as the executors of the historical process is not a permanent one.

That special piece of history called salvation history continually reminds us that we are not just created by God but toward God. Thus the unresolved contradictions of our historical existence will find their resolution in the larger context of salvation. While God in his general providence "provides both order and grace as the matrix of existence," (human) existence is not a self-contained phenomenon.[126] It is open toward its future which is foreshadowed in the Christ event as the promise and the proleptic anticipation of a new creation. Therefore we now turn to God's special providence.

---

125. M. Weinfeld, "berith," in *Theological Dictionary of the Old Testament*, 2:255f.
126. Outler, 52.

f. Special Providence

In our survey of God's general providence we have seen how God provides a trustworthy basis for human activity. In granting order to our existence God primarily sanctions natural constellations which already prevail in our world. Only in the eschatological provision of grace do we notice that God's providential activity will extend beyond the presently available. Does this mean that within the historical context we cannot expect anything new? That the phenomenon of novelty will only emerge when the eschaton commences?

We remember that in the evolutionary processes of life the elimination of the novel would not do justice to the historical facts. The evolution of life in its present form was not something simply to be expected, but it included both novelty and predictability. Turning to God's self-disclosure as it culminated in Jesus Christ, we notice too that there is constancy and surprise. God's involvement with humanity developed along certain lines (e.g., covenant, promise and fulfillment, law and gospel, etc.). At the same time, there are events that were totally unexpected, such as the crossing of the Red Sea, the election of David, and the coming of the Messiah. Even Christian existence stands under the same dialectic. On the one hand it is as predictable as any other human existence. But at the same time Christians have a vision of a new creation and are encouraged to proleptically anticipate something of the eschatologically envisioned goal. As Christians they enjoy already here and now an existence in the new creation. The eschaton as novelty is not just a future phenomenon, it is to some extent a present experience. To realistically affirm this eschatological novelty, we must understand that God's providence extends beyond the provision of order in his general providence. Just as God's self-disclosure is unexpected, so is God's special providence as God's provision of novelty in his miraculous activity. We must deal here primarily with the problem and significance of novelty, i.e., miracles, as well as with the issue of prayer as a response to and a request for God's providential care.

Miracles

In order to talk about novelty in terms of God's special providence, we should first clarify what we mean by this term. To merely replace the more traditional term "miracle" with the term "novelty" would not solve anything, since much of the problem involved in talking about miracles stems from the confusion about what a miracle is. Novelty or miracle is here theologically understood as an act of God which runs counter to our usual experience and which becomes visible in the objective world. A miracle is an exception. Something that occurs every day we usually do not call miracle or novelty. The feeding of the five thousand (John 6:1-14) has often been called a miracle. Augustine rightly commented: "For certainly the government of the whole world is a greater miracle than the satisfying of five thousand men with five loaves; and yet no man wonders at the former; but the latter men wonder at, not because it is greater, but because it is rare."[127] The problem posed by a novelty, however, is not its exceptional character, but that it becomes visible in the objective world. Thus miracles are erratic blocks and are in danger of becoming stumbling blocks. The reason for this is that as an act of God they pertain to the metaphysical dimension, while as something that has become visible they belong to the dimension of the natural or the physical.

Through the dominance of the natural sciences a more and more stringent distinction was made between the natural, the object matter of the natural sciences, and the supernatural, the realm of God. If God is to be understood as the agent of a miracle, one commonly assumed that he could not proceed in a natural manner but in a supernatural one. Thus a potential conflict was laid between those who felt that the natural context of events does not allow for divine disruptions and those who asserted that God can at any time interfere with the natural. It is interesting here to remember that at the time of Augustine, the distinction between the natural and the supernatural had not yet been made. He was still free to say: "A portent, therefore, happens not contrary to nature, but contrary to what we know as nature."[128] The strict distinction between the natural and the supernatural restricts — if not outright precludes — the possibility of a miracle.

It was only in the ninth century, after the Greek Neoplatonic works

127. Augustine, *Lectures or Tractates on the Gospel according to St. John* 24.1, in Nicene and Post-Nicene Fathers, 1st ser., 7:158.

128. Augustine, *The City of God* 21.8, in Nicene and Post-Nicene Fathers, 1st ser., 2:459.

of Dionysius (ca. sixth century), the Areopagite, an alleged student of Paul (Acts 17:17-34), had been translated into Latin, that the term "supernatural" finally made its appearance in Western theology.[129] Yet it was only after the thirteenth century, primarily with the help of Thomas Aquinas, that "supernatural" became a commonly accepted theological term. Thomas, for instance, clearly distinguished between the natural and the supernatural: "Firstly, there is natural change, which is done in the natural way by the appropriate agent. Secondly, there is miraculous change, which is done by a supernatural agent, above the normal order and course of nature — as for instance the raising of the dead to life."[130] Yet Thomas did not want to see the natural in opposition to the supernatural. Like most people in the Middle Ages, Thomas perceived God as the one who connects the natural with the supernatural so that the natural order is enveloped in the supernatural, which provides origin, sustenance, and a goal for the natural.[131]

Aquinas agreed with Augustine that a miracle is *"something difficult and unusual."* But then he added that it is also something *"surpassing the capabilities of nature [supra facultatem naturae]* and the expectations of those who wonder at it."[132] A miracle therefore is not simply something unusual and unexpected. It is something altogether wondrous, having its cause, namely, God, hidden absolutely. Since Thomas defined a miracle as an act of God which is outside the normal pattern of nature and surpasses its capabilities, he even went so far as to say: "Creation and the justifying of the sinner, while they are acts of God alone, are strictly speaking not miracles, because they are acts not meant to be accomplished by other causes." Thus a miracle, as an act of God, does not take place outside the natural realm but within it, replacing the natural course of things.

Such a definition rightly emphasizes that at certain points God's creative activity is present in nature in an unusual way. Yet the conclusion could easily be drawn that God is significantly less present in the usual proceedings of nature. Putting nature in opposition to the super-

129. For the following, cf. John P. Kenny, *The Supernatural: Medieval Theological Concepts to Modern* (Staten Island, N.Y.: Alba House, 1972), 94f.

130. Thomas Aquinas, *Summa Theologiae* 3a.13.2r; 49:159.

131. Cf. Thomas Aquinas, *Summa Theologiae* 1a.105.5r; 14:79.

132. For this and the following quotation, see Thomas Aquinas, *Summa Theologiae* 1a.105.7; 14:83ff.

natural, as had become more and more customary in medieval scholastic theology, was intended to emphasize the supremacy of God. God is above and beyond nature. By reserving the supernatural as God's unique and separate realm or mode of action, God was inadvertently placed into a rather aloof position. God usually let things run their course unless his special miraculous involvement was called forth. (The general acceptance of the distinction between natural and supernatural coincided with the growing significance and independence attributed to the natural sciences.) Thus unintentionally God was relegated more and more to the realm of the supernatural, and eventually he was completely divorced from the realm of the natural. Thus nature was gradually perceived as running independently according to its own laws. In the fifteenth century eminent scholars like Nicholas of Cusa still attempted to bridge the widening chasm between theology and science by assuming "the coincidence of opposites" in which all finite contradictions would merge into an infinite unity.[133] But his influence was not lasting; natural science dominated more and more, eventually excluding supernatural possibilities.

Until fairly recently Roman Catholic thinking followed the line of argument advanced by Thomas Aquinas. A miracle would be asserted if a certain phenomenon (for example, the healing of a sick person) could not be sufficiently explained by assuming only natural causes.[134] God's involvement in the context of a miracle was thus conceived as an action separate from the workings of other (natural) forces. However, two dangers could arise from this kind of thinking: (1) It could easily lead to the conclusion that God is not the sole agent of all processes. God interferes only at specific, unusual points through miraculous actions. (2) It might also make the occasions for divine interventions fewer and fewer, as our

---

133. Cf. Nicholas of Cusa, *The Vision of God* (9), in *Unity and Reform: Selected Writings of Nicholas de Casa,* ed. John P. Dolan (Notre Dame, Ind.: University Press, 1962), 149.

134. Cf. the extensive study by Louis Monden, *Signs and Wonders: A Study of the Miraculous Element in Religion* (New York: Desclee, 1966). Cf. also the five characteristics of a miraculous healing in Lourdes which were cited in Ruth Cranston, *The Miracle of Lourdes,* updated and expanded edition by the Medical Bureau of Lourdes (New York: Doubleday, Image Books, 1988), 125f. The five characteristics are: "1: Absence of curative agent (such as drugs or injections, special treatments, etc.), 2. Instantaneousness, 3. Suppression of convalescence, 4. Irregularity of the method of healing, 5. Function restored without action of the organ — still incapable of accomplishing it."

natural knowledge of the world increases. This latter point is demonstrated by the decreasing number of actual miracles officially admitted by the Roman Catholic Church. Gradually God becomes so transcendent that in more and more cases we only confront the natural events by themselves. There has become less of a need to resort to divine or miraculous explanations. God is then relegated to a supernatural sphere that has no bearing on our everyday life.

Protestant theology, however, has not fared much better than Roman Catholic theology. Some theologians attempted a distinction between the natural and the supernatural similar to Thomas's, and arrived at comparable results. Most theologians, however, in response to the rising dominance of science, attempted to "reconcile" the biblical miracles with our scientific knowledge. For instance, in the seventeenth century, when German Protestantism no longer found the worldview of Jesus tenable, the theory of accommodation was developed. This theory attempted to distinguish between the conceptuality with which Jesus proclaimed his message and the actual intent of his proclamation. Johann Salomo Semler (1725-91) further refined this theory in the eighteenth century. For example, he claimed that the Jews of Jesus' time believed that "all unusual and extraordinary physical evils were caused by evil spirits."[135] Though Jesus did not share this belief with his contemporaries, he accommodated himself to their thinking and, in the eyes of the Jews of his time, performed miraculous exorcisms. The intention of the "miraculous" expulsions of demons was to free these people from their fears.

A generation earlier the English Deist Thomas Woolston (1670-1733) claimed that the New Testament miracles could not have occurred in the way they were depicted by the Evangelists. If visualized, these events would lead to numerous contradictions.[136] Yet, like Semler, Woolston did not want to suggest that the New Testament miracles were fictitious. According to Woolston, some miracles, such as the story of the empty tomb, were most likely bare of historical content, but oth-

---

135. Cf. Anonymous, *Versuch einer biblischen Dämonologie oder Untersuchung der Lehre der heil. Schrift vom Teufel und seiner Macht, with a preface and an appendix by Johann Salomo Semler* (Halle: Carl Hermann Hemmerde, 1776), esp. 335f. and 341f., in Semler's appendix; quotation 87.

136. Cf. Emanuel Hirsch, *Geschichte der neueren evangelischen Theologie* (Gütersloh: Gerd Mohn, 1964), 1:316ff.

ers certainly did contain a historical kernel. Woolston, however, left open what this historical kernel was. He suggested that in the form in which they were told by the Evangelists, the New Testament miracles were at best only allegorical or mystical.

In the eighteenth century many natural explanations of the New Testament miracles were advanced. For instance, Jesus' walking on water acquired a natural explanation. He was merely wading in shallow water while his disciples believed he was actually walking on the water. The feeding of the five thousand was interpreted as the result of Jesus' powerful preaching. His audience was so fascinated by him that they forgot about their appetites. For a short while eating was secondary. Of course, not everyone was satisfied with such compromising explanations. To the dismay of Semler, Hermann Samuel Reimarus (1694-1768), for example, denounced the New Testament miracles as pious fraud deliberately invented by the authors. "Only thirty to sixty years after the death of Jesus, people appear who write down these miracles as if they happened in the world."[137] Our scientific knowledge of the world had now become the norm for what God could do. What was naturally possible according to the then available scientific knowledge was also regarded as supernaturally conceivable. Even in our days, Rudolf Bultmann (1884-1976) followed this line of thinking when he claimed: "It is impossible to use electric light and the wireless and to avail ourselves of modern medical and surgical discoveries, and at the same time to believe in the New Testament world of demons and spirits."[138]

In recent years, however, we have realized that the order and structure we discover in nature is not something that is there as a given, but it is partly something we introduce into nature. The English astronomer Sir Arthur S. Eddington (1882-1944) expressed this best when he said: "We have found that where science has progressed the farthest, the mind has but regained from nature that which the mind has put into nature. We have found a strange foot-print on the shores of the unknown. We have devised profound theories, one after another, to account for its

137. Hermann Samuel Reimarus, *The Goal of Jesus and His Disciples*, introduction and translation by George W. Buchanan (Leiden: E. J. Brill, 1970), 119.

138. So Rudolf Bultmann in his famous essay "New Testament and Mythology," in *Kerygma and Myth*, ed. Hans Werner Bartsch, trans. Reginald H. Fuller (London: SPCK, 1953), 5.

origin. At last, we have succeeded in reconstructing the creature that made the foot-print. And Lo! it is our own."[139]

We also remember that the so-called laws of nature are not laws according to which events must occur, though we usually expect this, but they are laws patterned according to our experience of the way events generally happen. Furthermore, we have noticed that the substructure of reality is undetermined. It allows for the kind of novelty that, for instance, characterizes the evolutionary process.

It would be wrong, however, to assume that our present understanding of nature allows for God's miraculous interference. The idea of God interfering with nature is also foreign to the biblical understanding of God's working. When we look to the biblical witness, we notice that God's miraculous activity is not viewed as something contrary to or superimposed on nature. "There is no talk of a sealed-in world or of iron-clad laws which must be broken through."[140] Since the biblical witness is convinced that God is involved in the totality of the world, miracles are viewed then as a new and surprising mode of God's ongoing activity. In other words, God's special providence is a peculiar but important case of his general providential activity. Paul Tillich (1886-1965) stated this succinctly: "Providence is not interference; it is creation. It uses all factors, both those given by freedom and those given by destiny, in creatively directing everything toward its fulfillment. . . . It is not an additional factor, a miraculous physical or mental interference in terms of supranaturalism."[141]

Arthur Peacocke writes:

> *Particular* events or cluster of events, whether natural, individual and personal, or social and historical, (a) can be specially and significantly revelatory of the presence of God and of the nature of his purposes to human beings; and (b) can be intentionally and specifically brought about by the interaction of God with the world in a top-down causative way . . . does not abrogate the scientifically observed relationships operating at the levels of the events in question. The combination of (a) and

139. Arthur S. Eddington, *Space, Time, and Gravitation: An Outline of the General Relativity Theory* (Cambridge: University Press, 1921), 200f.

140. Berkouwer, 222.

141. Paul Tillich, *Systematic Theology* (Chicago: University of Chicago Press, 1951), 1:267.

(b) renders the concept of God's special providential action intelligible and believable within the context of the perspective of the sciences.[142]

God's miraculous activity occurs within and through the present structure of nature. This does not imply, however, that a miracle as miracle becomes evident in the natural context. While God is equally and totally present to all times and places in nature and history, human awareness of that presence does not always have the same intensity. There are certain events that are more revealing of God's activity than others. Therefore a miracle has special revelatory significance. Yet this significance is not demonstrable. What becomes visible are two significant consequences of a miracle:

1. We observe that something ran counter to our usual sensory experience.[143] It may just be a striking constellation of causes, conforming with the laws of nature and occurring at the appropriate moment. For instance, such a constellation would be the strong east wind that according to Exodus 14:21 began to blow at exactly the right moment and enabled the Israelites to escape through the Red Sea while not allowing the pursuing Egyptians to take the same route. However, our observation that something ran counter to our usual sense experience could also be due to a special (causal) act that overrides the normal sequence of cause and effect in human affairs. An example for this kind of event would be the healing of the paralytic (Mark 2:11). Though medical science knows of exceptional instances in which paralyzed persons can regain the function of their limbs, there is no law that says that it will indeed be so. At most one can hope for such an event, but one cannot expect that it occurs. A miracle stands in contrast to other events that occur with regularity.

2. The other consequence of a miracle is that the miraculous event is perceived as an item of the past. Once the miracle had occurred, we would have seen that the Israelites had arrived at the other side of the sea and we would have watched how the formerly paralyzed person was walking again. Once these results are visible, we will have to enlist the help of all available expertise to bring the states prior to and after the miracle into a causal relationship. Being in charge of the natural context

142. Peacocke, 182.

143. See also Mark Pontifex, *Freedom and Providence* (New York: Hawthorn, 1960), 114.

in which we live, we will also want to find the natural causes that made the miraculous event possible.

When confronted with a miracle, our Christian faith will inform us that it was not just nature that made this "miraculous" change possible. It was, at the same time, the result of God's mighty hand through his special providential activity. Yet is such a twofold view of reality possible? Or does it simply give the same constellation a different name? It would be helpful here to remember the dilemma that scientists initially faced when they wanted to determine the nature of elementary particles. They were confronted with the seemingly exclusive duality between wave and corpuscle until the Danish physicist Niels Bohr introduced the principle of complementarity, suggesting that wave and particle properties are complementary aspects of a single reality. This principle allows for a twofold view of reality, a view which is not based on a temporary deadlock in scientific research but reflects "an essential characteristic of reality."[144] We should apply this insight to the binary components of a miracle, namely, the natural cause-and-effect sequence and God's miraculous activity of using these causes in an unusual or unprecedented way. We would conclude that though the visible presence of the one (e.g., nature) seems to exclude the presence of the other (e.g., God), both aspects, God and nature, complement each other and point to the whole of reality.

A miracle does not replace faith by demonstrating the presence of God through sign language. Rather it necessitates faith so that we allow for and affirm the total twofold reality that encounters us. William Pollard captured this situation very aptly when he said: "What to the faithful is an act of divine mercy showing forth our Lord's restorative power is for the pagan merely a piece of extraordinarily good luck. What to the faithful is a manifestation of divine judgment is to the pagan only a misfortune."[145] But with this evaluative remark we are already touching upon the significance of miracles for the process of salvation in general.

According to the biblical witness, miracles are intrinsically related to the process of salvation. Today even the most critical analysts of the

144. Pollard, *Chance and Providence*, 141; and cf. Günter Howe, "Zu den Äußerungen von Niels Bohr über religiöse Fragen," *Kerygma und Dogma* 4 (January 1958): 26f., where Howe points out the implications of Bohr's principle of complementarity for our understanding of reality.

145. Pollard, *Chance and Providence*, 66.

New Testament sources admit that Jesus did indeed perform acts that his contemporaries regarded as miraculous and that we still consider highly unusual.[146] Jesus accompanied his teaching ministry with a ministry of healing. He healed people such as the paralytic (Mark 2:11), Peter's mother-in-law (Mark 1:31), and people possessed by unclean spirits (Mark 1:26).

Jesus' miraculous activity is not without analogies. There are accounts of miracles performed by contemporaries of Jesus. These accounts show an astounding similarity to Jesus' own miracles. Apollonius of Tyana (ca. A.D. 3–ca. 97), for instance, an itinerant Neo-Pythagorean teacher and contemporary of Jesus, is supposed to have raised people from the dead and healed many who were sick.[147] Though some of the miracles may simply have been attributed to persons in antiquity to emphasize their importance, it would be overreacting to conclude that none of these miracles allegedly performed actually took place. Even the similar structure of miracle stories of the Gospels and of miracle stories in other literary sources of antiquity would not disprove their factuality.[148] There are not many variations possible in the way in which a miracle can be effectively told. Since, however, the power to perform miracles was at that point also considered a status symbol, symbolizing a special relationship to the gods, each claim to truth must be carefully analyzed. The possibility cannot be excluded that even some of the miracles attributed to Jesus are without historical basis, merely serving to underscore his exceptional status.

The New Testament writers did not seem threatened by the existence of other miracle stories. They were even convinced that the performance of miracles was not the exclusive prerogative of Jesus. They tell us that Jesus himself warned of false christs and prophets of the end times who would perform great signs and wonders (Mark 13:22). The apostles also knew of people who used sorcery to perform miraculous

146. Cf. Herbert Braun, *Jesus of Nazareth: The Man and His Time*, trans. Everett R. Kalin (Philadelphia: Fortress, 1979), 29.

147. Cf. Gerd Petzke, *Die Traditionen über Apollonius von Tyana und das Neue Testament* (Leiden: E. J. Brill, 1970), 125-37; cf. also Philostratus, *The Life of Apollonius of Tyana* (4.6), trans. F. C. Conybeare (Cambridge: Harvard University Press, 1969), 1:367.

148. Rudolf Bultmann, *The History of the Synoptic Tradition*, trans. J. Marsh, rev. ed. (New York: Harper & Row, 1968), 210 and 231ff., who tries to explain their similar structure through the hypothesis of a common origin.

deeds (cf. Acts 8:9). If we want to obtain a complete picture of the significance of the miracles of Jesus, it is insufficient to interpret them exclusively in the light of Near Eastern or Greco-Roman religious thought.

Since Jesus' miracles were part and parcel with his mission, we must understand them primarily from the perspective of the purpose and destiny of his life. If we can agree with the New Testament writers that Jesus announced the kingdom of God and brought it about through his life and destiny (Mark 1:14f.), then Jesus "was creating the future by his wonders; the forces of the world to come were already being manifested in and by him."[149] Miracles are therefore never used by Jesus to demonstrate his power and to legitimate himself (cf. Matt. 12:38f.). They are rather signs that illustrate his message. They show that God is not a distant and far-off God, but a God actively involved in the creative process. God does not simply confirm present tendencies, he is willing to give them a new and unprecedented turn. This creative activity of God results in a decisive confrontation with the antigodly powers. According to the New Testament writers, Jesus did not perform his miracles aloof and detached from affairs of the day. Authorized by God, he fought and overcame destructive antigodly powers, and each of his victories then became visible in a miracle. In a vivid and dramatic way the Evangelists tell us that these powers recognized Jesus and exclaimed in anguish: "What have you to do with me, Jesus, Son of the Most High God? I adjure you by God, do not torment me" (Mark 5:7).

We have seen that God fends off the destructive antigodly powers through his orders of preservation in natural, moral, and historical processes. Through his miracles, however, he does not just maintain order, but, in a creative act, initiates a completely new order. The antigodly powers are not merely kept in check, but at one specific point they are overcome, they have to retreat. Therefore miracles are signs of the commencing kingdom and reign of God.[150] We obtain a glimpse of an en-

---

149. Anton Friedrichsen, *The Problem of Miracle in Primitive Christianity*, trans. Roy Harrisville and John S. Hanson (Minneapolis: Augsburg, 1972), 73. Cf. also Rudolf Bultmann, *Theology of the New Testament*, trans. Kendrik Grobel (New York: Scribner, 1954), 1:7, who shows that Jesus sees the fulfillment of the prophetic predictions of salvation "already beginning in his own miracles."

150. Cf. Joachim Jeremias, *New Testament Theology: The Proclamation of Jesus*, trans. John Bowden (New York: Scribner, 1971), 95, who rightly claims that the victories over the power of Satan "are a foretaste of the eschaton."

tirely new creation, when sick people are restored to health, the dead are brought back to life, biologically impairing phenomena (such as hunger) are overcome, and physically limiting phenomena (such as space and gravity) are eliminated. The seer in Revelation, envisioning the eschatological perfection, conveys a similar picture when he says:

> Death will be no more;
> mourning and crying and pain will be no more,
> for the first things have passed away. (Rev. 21:4)

Of course, Jesus' miracles are only temporary points of victory over the antigodly powers. Those who are healed may become sick again, those who are brought back to life will die again, and those who are fed will once again be hungry. Does this mean that the present structure of reality is so overpowering that even miracles would at best provide a temporary escape and not an indication that the present structure of reality will be taken up into a new structure, the new world to come? We could only answer this question in the affirmative if we were to neglect the resurrection of Jesus Christ as the miracle through which all other miracles of Jesus are endowed with ultimate validity. Jesus' resurrection was not a resuscitation or a return to life after which another and final death followed. As Paul victoriously exclaimed: In Jesus' resurrection death was overcome through a new form of life (1 Cor. 15:55ff.). Here the promised transformation of the whole cosmos had commenced.

The reality shown to us in the resurrection of Jesus Christ was not a restoration of our present cosmos, it was an indication and anticipation of a new cosmos.[151] Since this miracle, the inauguration of a new reality, happened with and because of Jesus, we are allowed to accept all the miracles that he himself performed and all miracles that are still performed in his authority as signposts foreshadowing and pointing to a new world to come. Miracles have eschatological significance. They point to the promised eschaton and anticipate it proleptically. They indicate that our present world is not endowed with permanence but is on course to a new world. Through his special providence God reminds us that the orders of preservation are just that, orders that preserve the

---

151. Cf. William Manson, "Eschatology in the New Testament," in *Eschatology: Four Papers Read to the Society for the Study of Theology,* ed. William Manson et al. (Edinburgh: Oliver and Boyd, 1952), 6.

world for its fulfillment and perfection in the new world to come. In his special providence God also shows us that these orders are only of penultimate quality. Though they are usually reliable, they are not so restrictive as to exclude novelty, even in the sense of the ultimate and universal novelty of the new creation.

We must refrain, however, from the utopian assumption that there is a developmental continuity between our present structure of reality and salvation in and through Christ. As we have learned from Teilhard de Chardin, evolutionary pressure provides, at best, the elements for the Christogenesis, but by itself this pressure will not bring the Christogenesis about. To achieve salvation, neither evolution nor revolution will suffice. Salvation can only be brought about by something unprecedented and new, by the creative activity of God as foreshadowed in the Christ event.

It should be noted, at least parenthetically, that a miracle taken by itself does not necessarily have convincing power. It does not necessitate the conclusion that we are confronted here with God's salvific activity. For instance, many of those who saw that the sick were restored to health were amazed and glorified God (Mark 2:12). Even Jesus' "friends" and the scribes did not deny that Jesus had performed unusual deeds. But they concluded that he was either "beside himself" or was connected with "Beelzebul, and by the ruler of the demons he casts out demons" (Mark 3:21f.). A miraculous act by itself is silent. It does not disclose whether it (1) is endowed by God with eschatological significance, (2) is a sign that we are confronted with an especially gifted person, such as a true miracle worker, or (3) is a seductive act of antigodly powers. For Jesus, however, sign and proclamation go together. As we can see with his "friends" and the scribes, those who rejected his message did not change their minds once they were confronted with his miracles. They had heard the miraculous message of the commencement of the salvational process and rejected it together with its signs. Here the conclusion of the parable of the rich man and Lazarus conveys a most telling insight: "If they do not listen to Moses and the prophets, neither will they be convinced even if someone rises from the dead" (Luke 16:31).

Prayer

In our considerations of God's care for humanity we did not want to leave the impression that God's preserving and promising activity relegates us to inactivity. If such were the case, it would violate our position as God's governors in the world. Once we turn to the impact of prayer on God's providential care, we will soon notice that we are encouraged to be actively involved in this providential work. We could even say that the experience of God who hears and answers our prayers is at the heart of the question of providence.[152] If God did not interact with us in a dialogical way, we would be confronted with an impersonal "it," as in the Stoic worldview, with the laws of nature or with a merciless fate. Prayer would be nothing but an attempt to calm our nerves. It would just be an attempt to obtain self-control, analogous to Far Eastern meditation exercises.

But the New Testament is full of exhortations to pray which leave no doubt that an actual I-Thou relationship between God and humanity is envisioned (cf. Matt. 7:7; 21:22; John 15:7). Consequently we are told that one does not address God with a carefree, casual attitude. Neither should one pray in a boastful manner, convinced that God has no choice but to agree with the contents of our prayers (Matt. 6:5-7). Instead a prayer should be precise and made in humility. It is not without significance that in the Lord's Prayer (the prayer Jesus taught his disciples) the assertion of the holiness of God's name comes first (Matt. 6:9).[153]

This emphasis on God's holiness informs not only the attitude of our prayers but also their content. For instance, according to the Gospel of John, Christ promises, "The Father will give you whatever you ask him in my name" (15:16). This promise does not imply that God will grant us anything for which we ask him. Dietrich Bonhoeffer captured the meaning of prayer well when he said: "God does not give us everything we want, but he does fulfill all his promises."[154] Prayer is not a friv-

152. Cf. Georgia Harkness, *The Providence of God* (New York: Abingdon, 1960), 121, who also reminds us that this is the place at which many people's faith in providence is shipwrecked.

153. Cf. Joachim Jeremias, *The Lord's Prayer,* trans. John Reumann (Philadelphia: Fortress, Facet Books, 1969), 21f.

154. Bonhoeffer, *Letters and Papers,* 206, in a letter of August 14, 1944.

olous attempt to discover how far-reaching God's power is. For example, one should not ask God to reverse the sequence of winter and spring in order to see if he can change the order of the seasons. Rather the intent is to rely, for Christ's sake, on the promise expressed in the Psalms:

Call on me in the day of trouble;
  I will deliver you, and you shall glorify me. (Ps. 50:15)

Since Christ has overcome the destructive antigodly powers, we, as his followers, are encouraged to walk beside him and call upon God to deliver us or others from the impact of these powers.

Martin Luther was right when he said that "God's order or command and the prayers of Christians . . . are the two pillars that support the entire world" and without which the world would disintegrate.[155] God promises to consider the content of our prayers in his preserving, sustaining, and creative activity. Through our prayers we express our solidarity with God, cooperating with him and dialoguing with him concerning the future of the world. Prayer therefore can have many purposes. Its content will express these purposes. First of all we must mention adoration and praise of God. Again Luther reminds us that we should not only call upon God in our plight. We should also thank him for his help and rescue, remember his acts of kindness, and praise him for them, because "He is the Creator, the Benefactor, the Promiser, and the Savior."[156] Luther can rightly say that it is sinful if we cease to pray to God. A life without prayer is no longer in tune with God as the creative source of all life; it mistakenly presumes that our world is self-sufficient. Prayer serves here as a reminder for us to recall the one from whom we have everything that is. Therefore we should ask God even for things that we seemingly take for granted, such as good weather or a good harvest.[157] With this last sentence we have already touched upon the large category of petitions which seem to form the content of most prayers.

Luther also encourages us to bring before God all our anxieties,

---

155. Martin Luther, *Sermons on the Gospel of St. John* (1537/38), in *LW,* 24:81, in his explanation of John 14:12.

156. Martin Luther, *Lectures on Genesis (1534-45),* in *LW,* 3:117, in his explanation of Gen. 17:7.

157. Cf. Martin Luther, WA, 37:425.2-8, in a sermon on Ps. 65 (1534).

such as personal afflictions like poverty and sickness, or even sinfulness. He asserts that we should not exclude any petitions, whether they envision "temporal or spiritual things."[158] Georgia Harkness (1891-1974) rightly prioritizes this all-inclusive scope of petitions when she says that "to seek God's forgiveness for past and present sin, and thus to find hope for the future, is an essential part of Christian prayer."[159] If we do not include in our prayers the plea for forgiveness of sin, our dialogue with God will always be disturbed. Both in our prayers and in our expectations we would act out of our own sinful and selfish interests and not out of conformity with God. Petitions are therefore concerned first with inner strength and renewal. Of course, petitions will include more than asking for forgiveness. They will also be prayers for inner peace in times of conflict, for clarity of outlook, for new strength at moments of fatigue, and for power to cope with the daily demands of life. The following stanza from the gospel hymn "What a Friend We Have in Jesus" expresses this kind of prayerful attitude very appropriately:

> Have we trials and temptations?
> Is there trouble anywhere?
> We should never be discouraged —
> Take it to the Lord in prayer.
> Can we find a friend so faithful
> Who will all our sorrows share?
> Jesus knows our ev'ry weakness —
> Take it to the Lord in prayer.[160]

From the acknowledgment of God's benevolent activity in Jesus Christ, new strength and peace of mind are gained. God as the ruler of the universe does care about us little, unimportant beings. He cares so much that he has come to us in the human form of Jesus Christ. This God who cares is also the one who gives strength to the weary and lifts up those in low esteem (Luke 1:52).

In our consideration of petitionary prayer, we dare not forget the frequent petitions for physical health and healing. Since prayers are not

---

158. Martin Luther, *Instructions for the Visitors of Parish Pastors in Electoral Saxony,* in *LW,* 40:279.

159. Harkness, 128.

160. Hymn 439 of the *Lutheran Book of Worship.*

intended to be a substitute for work, petitions for recovery from physical illness should never replace appropriate medical care. However, the two are not mutually exclusive. The same insight must guide our attitude toward so-called faith healing. Though each case of a miraculous healing must be subjected to careful scrutiny, we know that unusual and unforeseen recoveries from grave illness do occur. There is also no doubt that some people have the gift of healing.[161] We have seen, however, that this power need not stem from God. It could also be obtained from antigodly, seductive powers.

Again we are confronted with the fact that a miraculous event is silent, it does not disclose its originator. However, in prayer we have a means with which we can "discern the spirits."[162] If we are existentially involved in an event of so-called faith healing or in any other miraculous event, we are able to discern the source of this healing power. The existential involvement will usually assume one of three forms: (1) we may be the one who has the gift to heal; (2) we may be the one who has been healed; or (3) we may be an immediately involved bystander (relative). If our relationship with God in prayer is strengthened, we may safely assume that the healing power was a gift of God. It will then not become commercialized or used for self-promotion and self-exaltation. Following the example of Jesus, it will rather be used to illustrate the Christian gospel and humbly and gratefully to further the human good. If, however, the relationship with God in prayer is weakened, we can hardly attribute this power to God. Its source must be antigodly powers who enable seemingly divine miracles to seduce people, often even accompanied externally with Christian symbols. But if the relationship in prayer remains unaffected, we may simply regard this healing power as an unusual, "natural" gift of God, similar to the superior healing gifts of some medical doctors.

In this new century we have more and more come to realize that a human being is a psychosomatic unity. We notice that psychic disturbances, such as depressions and neuroses, can bring about physical ailments, problems with the digestive system, malfunctioning of the glands,

161. Cf. also the entries: "Healing, Psychic," "Healing by Faith," and "Healing by Touch," in *Encyclopedia of Occultism and Parapsychology*, ed. Leslie A. Shepard, 2nd ed., 3 vols. (Detroit: Gale Research, 1984), 2:596-602.

162. This has been pointed out very convincingly by Karl Heim, *The Transformation of the Scientific World View*, trans. W. A. Whitehouse (London: SCM Press, 1953), 192f.

and heart and kidney diseases.[163] Psychic disturbances are frequently intertwined with spiritual crises. Regaining psychic and spiritual strength and balance is often accompanied by a physical healing process. The strengthening, alleviating, and comforting impact of prayer cannot be underestimated. Prayer can indeed be effectively "used" to calm our nerves. Such use, however, does not result in a dialogue with God but moves on the meditative level of our own psyche. We should refrain from calling it prayer, but term it more appropriately meditation. Though it dare not become a substitute for prayer, it serves a rather useful function in our turbulent times. It can help one attain a state of tranquillity and peacefulness of mind. Of course, we should not expect that the dialogue with God which is found in true prayer is followed by an automatic physical improvement. Restoration of psychic and spiritual health is usually a very slow process and occasionally will not be attained at all. If we take seriously the dialogical character of prayer, we must also be ready for God's noncompliance with our petitions.

When we come to petitions concerning natural events, we must bear in mind that this is primarily the realm where we will affirm the natural protective divine orders. Yet it is part of our task as God's administrators to remind him of those who are especially exposed to the dangers of the forces of nature, such as miners, travelers, pilots, and sailors. Again prayers are not intended to replace protective measures, but to accompany and perhaps enhance them. Similarly it is our prerogative and duty to pray in adverse conditions, such as storms, floods, and other disasters, that their impact will be softened or averted. Since a prayer is never uttered in selfish interest, we will keep in mind not only the well-being of ourselves but of others as well. This means the same adoration, praise, and petitions that we extend on our behalf, we extend also on behalf of others.

In all our prayers we always conclude with the expressed or tacit admission that God's will — not ours — be done. In Gethsemane Jesus prayed what is the most fervent prayer ever uttered: "Yet not what I want but what you want" (Matt. 26:39). So should we likewise pray. A Christian prayer is not a demand for God's surrender, but rather the prerogative and duty of a dialogue with the one who has formed the earth and

163. Cf. Harkness, 135ff.; see also Paul Tournier, *The Healing of Persons*, trans. E. Hudson (New York: Harper, 1965).

the whole universe, and who has been our dwelling place in all generations (Ps. 90:1f.).

When we reflect once more on divine providence, it should be clear that God cannot be understood as the primary cause that works in and through all secondary causes in nature and history.[164] If it were so, humans would only execute God's will without having any responsibility of their own. We cannot resort to two different perspectives or languages, the language of the natural sciences and the language of faith. Of course, there is always the possibility of resorting to a scientific explanation which does not take into consideration God's existence and activity and the theological interpretation which explicitly thematizes God's existence and working. Yet many items which are expressed scientifically do not even call for a theological interpretation, so that both ways of perceiving reality are not exactly parallel to each other. Perhaps one could say that God acts in all events through influence or persuasion. But again there are many events that have nothing to do with God's activity even though they do not occur without God's knowledge. For instance, when one person murders another, then this cruel event cannot be traced back to God's influence or persuasion. Here destructive and antigodly powers are at work.

To do justice to God's providence we must first recognize that through all the accidents, through all the different contingent events, ultimately something develops which in retrospect can be seen as the result of divine providence.[165] There emerges a certain order which allows life to originate and to develop. Therefore we cannot agree with Jacques Monod when he writes that "a *totally* blind process can by definition lead to anything; it can even lead to vision itself."[166] Monod, however, concedes that once chance has worked its effect, it is bound to that which it brought forth. Therefore one cannot deny the possibility "that there is a divinely ordained general direction, in which the process of the world is moving."[167] Though there existed no necessity that the

164. Cf. Owen C. Thomas, ed., *God's Activity in the World: The Contemporary Problem* (Chico, Calif.: Scholars Press, 1983), 231ff., who, at the end of this collection of papers, lists five different ways of God's activity in the world.

165. Cf. for this D. J. Bartholomew, *God of Chance* (London: SCM Press, 1984), 143.

166. Jacques Monod, *Chance and Necessity: An Essay on the Natural Philosophy of Modern Biology*, trans. Austryn Wainhouse (New York: Alfred A. Knopf, 1971), 98.

167. John Polkinghorne, *Science and Providence: God's Interaction with the World* (Boston: Shabhala, 1989), 40.

world developed in such a way as it shows itself now, it is evident that it has developed itself in such a way and therefore made life possible. The question is unimportant whether God was more "in a narrow curve" or on the "long stretches" of the evolutionary process. We should affirm with the Priestly creation narrative that it was not chance and evolution which brought us about, but God's will. Without God there would not have been possible an initial contingent event nor any other singular development. God is the one who is behind everything and who made possible the whole process. In a comprehensive way God is the creator and sustainer of everything that is. God is synonymous with life and its preservation.

But how shall we regard the hindrance or destruction of that which is, be it through natural catastrophes, death, or human interference? If God counteracts these destructive tendencies in his special providence, we cannot talk about this providence in a selective or manipulative way. For instance, if in a railroad accident I were "miraculously" spared but others were not, then it borders on blasphemy to talk about providence. With the same assertion we would remove from God's providence those who died. The confession that God wants our best does not mean that in every moment the optimally best is occurring. We are not allowed to dissect history atomistically in its individual components and interpret them without regard to the overall picture even if we are often tempted to do this. History only makes sense in its unity, because we know and confess its eventual direction, the new creation and the kingdom of God. Up to that final point, however, it is always endangered by events of nature, by human malice, and by plain human incompetence and negligence. The whole can only be interpreted appropriately from its end — this means eschatologically.

But where is God in the individual events? After God has appointed us humans to be his cocreators and representatives, God will not treat us as puppets. We are free to shape the course of history largely in an autonomous way. Yet God has not retired from history. In his general providence God grants stability and preservation. In this preserving activity God can also interact in a special way, as the human experience of God's special providence shows. The secondary causes are not simply suspended, so that God would interfere in the world machinery as an external force. In most cases it is sufficient that the appropriate constellations occur at the right time and place. "The Christian understanding of providence steers a course between a facile optimism and a fatalistic pessimism."[168] As Christians, we can trust that God will ultimately have everything in his hands, though we know he will not frivolously change the process of nature and history. The comprehensive change of nature and history is only envisioned at the completion of creation.

g. Completion of Creation

Our consideration of the completion of creation will also be done in the context of the natural sciences.[169] But we must keep in mind that eschatology is not the same as futurology. We can neither present a schedule of the so-called final events nor can we show in detail what the completed creation will be like. Predictions in this respect would presuppose a continuity between the present state of creation and its completion, whereas from a scientific and theological angle we must emphasize discontinuity.

Discontinuity must be a leading concept with regard to the completion of creation, because according to our present knowledge our cosmos is limited in space and time. This means that the energy available in the cosmos is limited. At some point life in our universe, as we experience it now, will come to an end. This insight is reinforced by the first two laws of thermodynamics. The first law, or the law of the conservation of energy, was given its most elegant formulation by Hermann von Helmholtz (1821-94) when he wrote: "The quantity of all forces which can be put into action in the whole of nature is unchangeable and can neither be increased nor decreased."[170] This statement must be seen together with the second law of thermodynamics, or the law of entropy. Rudolf Clausius (1822-88) stated this law in this manner: "The energy of the universe is constant. The entropy of the universe tends toward a maximum." This means that the quantity of energy in our universe is unchangeable, while the different levels of energy will come ever closer

---

168. Polkinghorne, *Science and Providence,* 44.

169. Concerning eschatology in general, cf. Hans Schwarz, *Eschatology* (Grand Rapids: Eerdmans, 2000).

170. Cf. Erwin N. Hiebert, "Modern Physics and Modern Faith," in *God and Nature,* 425, for the formulation of the laws of thermodynamics and the caution against the misuse in their interpretation.

to each other until the nonconvertibility of energy reaches its maximum. The implication is that someday in the future "the lights will be switched off" in the universe. On a smaller scale we encounter this already on earth when we notice that its natural resources are limited and that we are threatened by their depletion on account of our continuously increasing use of them.

Of course, there have often been attempts to escape the fate of an aging and finally dying universe. Pierre Teilhard de Chardin, for instance, claimed that the law of entropy applies to inanimate nature, but life is exempt from it.[171] With this statement, however, he has disregarded that animate nature can only live on the basis of inanimate nature. When our resources of carbon-based fuels on earth are depleted and when the sun ceases to shine, then all the energy resources will have been used up. There is no longer any possibility for life to continue. We are confronted then with the fact of the so-called heat death. Of course, this kind of scenario is still billions of years away so that we need not worry about this ultimate destiny. Much closer to us, however, is the so-called ecological catastrophe, as it is called by many people. The plundering of the natural resources, the seemingly unlimited increase in population, and the immense pollution of the atmosphere and of our environment in general could easily make the earth a place no longer fit for human habitation. This could happen in the foreseeable future. Already before one became aware of this possible ecological catastrophe, some people envisioned a nuclear holocaust. This disaster could occur through an atomic war or huge nuclear accidents. Then our earth would be uninhabitable and it would spell the end for humanity.

Teilhard, however, objected to such scenarios. These possible catastrophes would only occur within our universe but would not imply the end of the universe in its totality.[172] Even as a wreck the universe would still be around. These scenarios do not have eschatological relevancy, since they do not envision the end in an ultimate sense. Moreover, the Christian understanding of the end does not imply a final catastrophe. We cannot equate it with negativity, although in apocalyptic thinking

171. So Pierre Teilhard de Chardin, *The Vision of the Past*, trans. J. M. Cohen (New York: Harper, 1966), 168ff.
172. See Pierre Teilhard de Chardin, *The Future of Man*, trans. Norman Denny (New York: Harper Torchbook, 1969), 321f.

the final events are always accompanied by catastrophes. According to the Christian understanding, at the end there is a new creation.

When some conservative Christian circles suggest that an ecological catastrophe would entail the salvation of humanity through the return of Christ, such a speculative projection is not in line with New Testament assertions. In scientific cosmological perspective a completion of creation in terms of a new creation out of annihilation would only be possible if we would live in a pulsating universe, because then there is complete discontinuity from one cycle to another. Yet it is unlikely that our universe is a pulsating one. Moreover, a pulsating universe would always suggest that there are further cycles. Any completion would only be a transitory moment, and it would not be endowed with finality. The conviction of the biblical authors, however, is that the completion of creation will not be replaced by further creative acts. Completion means that everything is such as it ought to be, that beyond that completed state there are no further developments. From a scientific perspective, nothing can be said about the completion of creation. From that angle we can only receive the certainty that our present world will someday meet its death.

Hope and the expectation of something new has a long history in the Bible. The latest date of its emergence in the Scriptures is in Deutero-Isaiah. Here Yahweh announces:

> Do not remember the former things,
>   or consider the things of old.
> I am about to do a new thing. (Isa. 43:18f.)

Here the term *bara* (to create) is used to point out that there is a divine creative activity analogous to the creation in the beginning. But the primeval period is not equated with the end times, so that the creation in the beginning and at the end would correspond with each other. According to Deutero-Isaiah, the new creation is not a repetition of the creation in the beginning; it is something "radically new."[173] This announcement of a new creation is picked up in the New Testament by Paul when he says: "Everything old has passed away; see, everything has become new!" (2 Cor. 5:17). The end time is seen here in analogy to the

---

173. So Bernhard W. Anderson, *Creation versus Chaos: The Reinterpretation of the Mythical Symbolism in the Bible* (Philadelphia: Fortress, 1987), 131.

primeval time of creation. Yet it is no repetition of the beginning, nor a cyclic return to it; it is the fulfillment and high point of God's creative activity.

In the apocalyptic period this typology between primeval time and end time is transcended, and one understands history as a meaningful drama which will certainly fulfill God's plan which he had conceived already at the beginning. History is understood in a universal way. It is as a historical and cosmic drama which moves from beginning to end. History will be completely transformed into a new creation, and a new heaven and a new earth will emerge. There the metaphor of chaos is used in this context to show that the whole drama of salvation is accompanied from beginning to end by a cosmic rebellion until God finally overcomes the antigodly destructive powers. God's creation is continuously threatened by chaos and destructive powers until God accomplishes his plan in a universal manner.[174] Certainly human desires, anxieties, and dreams have influenced this apocalyptic understanding of the world and its ultimate destiny.[175] But it would be dangerous if we were to forsake the basic structures of this worldview, because then the cosmic breadth for a new creation would be missing. We would withdraw to the existential self of humanity instead of remembering that God has not created humanity just with body and soul but also with a cosmos.

Today the word of the apostle Paul about the "eager longing" of creation (Rom. 8:19ff.) is viewed as the hermeneutical key for the doctrine of creation.[176] Paul's statement does not just show the connection between humanity and the world, but also that the new creation must be seen in its cosmic dimension. The completion of creation is not restricted to humanity, but "Paul views the whole extra-human creation which surrounds a renewed humanity."[177] Paul picks up the Old Testament understanding of salvation as, for instance, indicated by Deutero-

---

174. Cf. Anderson, 143.

175. Cf. Christian Link, *Schöpfungstheologie angesichts der Herausforderungen des 20. Jahrhunderts* (Gütersloh: Gerd Mohn, 1991), 2:589, who states: "We have to abandon the world-view of the apocalyptic." Jürgen Moltmann takes an opposing position. Cf. his *Theology of Hope: On the Ground and the Implication of a Christian Eschatology,* trans. James L. Leitch (Minneapolis: Fortress, 1993), 136f., where he emphasizes the positive (cosmic) aspects of the apocalyptic.

176. So also Link, 2:387.

177. Alexandre Ganoczy, *Schöpfungslehre,* 2nd ed. (Düsseldorf: Patmos, 1987), 77.

Isaiah, and emphasizes that the whole environment, the whole cosmos, is supposed to become the context for salvation history and will reap the benefits of the salvation of the new humanity. Interestingly, in talking about the whole creation Paul uses apocalyptic metaphors "depicting creation as a person waiting for a momentous occasion," its own transformation.[178] He points to a creation which, in its chaotic state, is entangled in a cosmic struggle and is moving to the same goal as humanity. Yet creation has not reached its goal, because chaos, transitoriness, and destruction still characterize that which once was created as God's good creation. Paul, however, is convinced that creation will be liberated from enslavement to these destructive forces. It will not only be a spectator in the liberation of humanity, but creation itself will participate in this new state of being.

Why is Paul so optimistic that, influenced by apocalyptic literature, he talks about a new creation (cf. 2 Cor. 5:17)?[179] Paul does not talk about a restitution of creation, but of establishing a completely new creation. Through God's power this totally new reality will be and is already present and active through the gift of the Spirit. This Spirit, the creative power of God, can be understood as a "down payment" for the future salvation. We can already feel something of this new creation in the present if we are "in Christ."[180] In Christ, in his sphere of influence, we participate in the new creation and are liberated from the threatening and limiting forces of the present world. "Being in Christ" is understood as a mystic union with him. It does not mean that we simply structure our life in the way Jesus showed us when he lived among us as a human being. It means rather that we are closely connected with that which, with reference to the Old Testament, Paul expresses in the following way: "'Death has been swallowed up in victory.' 'Where, O death is your victory? Where, O death, is your sting?' The sting of death is sin, and the power of sin is the law. But thanks be to God, who gives us the victory through our Lord Jesus Christ" (1 Cor. 15:55ff.). Hope in a new creation and participating in it in a proleptic way is possible only be-

178. Joseph A. Fitzmyer, *Romans,* Anchor Bible Commentary (New York: Doubleday, 1993), 506, in his exegesis of Rom. 8:18ff.

179. Victor Paul Furnish, *II Corinthians,* Anchor Bible Commentary (New York: Doubleday, 1984), 314, in his exegesis of 2 Cor. 5:11-19, points out that the expression *kaine ktisis* (new creation) stems from apocalyptic Judaism.

180. Cf. Furnish, 332f.

cause this new creation has already occurred at one point in Jesus Christ. He is, so to speak, the "pledge and beginning of the perfect fulfilment of the world, as representative of the new cosmos, as dispenser of the Spirit, head of the church."[181]

In the resurrection of Jesus Christ a new creation has occurred for the first time. Jesus was not snatched away from death and brought back to life, but as the Christ he received a totally new way of existence which was no longer subjected to the laws of our world. This new creation, however, was no private act which was, so to speak, Jesus' own privilege. Since Jesus is the Christ, this new creation has universal significance and those who confess allegiance to him can also hope for a new creation. Moreover, they can hope with Paul that this new creation will extend to the whole cosmos. As the resurrection of Jesus Christ occurred in our world (though it was restricted to the destiny of Jesus), so the new creation in its fullness at the "day of the Lord" (1 Thess. 5:2) will again occur in this world and the Last Day will indeed be the last day of this world.[182] The day of the Lord is not restricted to Jesus Christ (i.e., to the destiny of an individual), but it has universal significance: therefore this Last Day will be the last day of the old world and at the same time the first and everlasting day of the new world. If we were to separate this day from our chronology, we would have to assume that the history of our world would continue indefinitely. But it is the peculiarity of a cosmic hope that the whole world will be drawn into an entirely new future. "The faith in a new creation cannot do without a christological foundation."[183]

When we affirm the cosmic dimension of the new creation, we cannot just look at the cosmological discoveries within the last century. We should also consider what the Old and the New Testament understood by cosmos. It was certainly not a universe that extended billions of light-years. Rather it was the cosmos that was known to them, like the sun, the moon, and some planets, such as Venus, the morning star. While those of us who live in the twenty-first century have a much broader knowledge

---

181. Karl Rahner, "Resurrection," in *Sacramentum Mundi: An Encyclopedia of Theology,* 5:333.

182. The contrast between "outside" and "inside" and between "in our time" and "at our time" (cf. Link, 588 and 593) does not make much sense here. What happens inside is also important for the whole (i.e., from outside), and vice versa.

183. Link, 591.

of the universe, we are no more able than our ancestors to interact with the depth of the universe. Interstellar or intergalactic journeys are, as far as we can see, only part of science fiction. For the foreseeable future even a mission to Mars by people rather than robots does not seem to be in the offing.

While we cannot exclude the possibility of sentient life in other regions of the universe, any communication or even visitation seems to be beyond our reach. To accomplish such communication, we would not only need to overcome the immensity of space, but we would also have to presuppose that the development on earth and in these other parts of the universe has occurred in exactly parallel ways. If we were only off by one hundred years, we would not yet have electronic gadgets even to attempt communication with distant regions. Even if they had such instruments, our receptors would most likely be missing and severely hamper any significant interaction.

To make a difference to us the creation of a new heaven and a new earth must include at least a cosmic change within our region of the universe. We know from cosmological observations that, while the overall picture of the universe is uniform, different regions are at different developmental stages. Therefore a change in one part of the universe need not necessitate a change in the whole universe. We could still hope for a new creation of our perceived heaven and earth even if the development in distant regions of the universe would run differently. We could similarly conjecture that if there are indeed other sentient beings in other parts of the universe, their redemption may take an analogous course to our redemption but not a synchronized one. Yet any hope we have for the future beyond the eventual rundown of the universe will not occur without God, our creator, sustainer, and redeemer.

While we appreciate the speculations of Frank Tipler and others, that intelligent life, in the sense of an information-processing capacity, might conceivably survive once we have depleted our resources on earth and even beyond, these visions do not entail much hope. First of all, science does not give us any reason to conclude that intelligent life must survive.[184] Scientifically speaking, we know with virtual certainty that our earth will disappear in approximately 4 billion years when the sun runs out of fuel. We can also conjecture possible destinies for the

---

184. Cf. for the following, Peacocke, 345.

universe as a whole. One possibility is a continuous expansion to an eventual heat-death when all energy levels will come so close to each other that the nonconvertibility of energy reaches its highest level. The other possibility is a slowing down of the expansion of the universe followed by a contraction or a "big crunch" and then possible further cycles of expansion and contraction in some kind of pulsating fashion. Neither one of these scenarios entails the necessity for life to survive. If Tipler were indeed correct in his speculations — speculations which must be based on other grounding than that of science — everything ever thought and done by human beings would be preserved in computerlike fashion. This grand summary would include all evils of the past and inhuman acts and therefore would not lead to a meaningful hope. Most likely good deeds and evil acts would cancel each other out.

Christian hope is based neither on science nor on the denial of science, but on God's self-disclosure as it showed itself in the life and destiny of Jesus the Christ. We can justifiably hope for a new creation, since it has been proleptically anticipated in the resurrection of Jesus Christ. Yet this new creation has a cosmic dimension. The seer of the book of Revelation does not only mention that God will

> wipe every tear from their eyes.
> Death will be no more;
> mourning and crying and pain will be no more. (21:4)

But he also speaks about a new Jerusalem (21:10) and says the people will bring into the new Jerusalem "the glory and the honor of the nations" (21:26). But before that can occur, all antigodly destructive forces must be overcome so that God can realize his kingdom in an unlimited way (19:6) and a new creation can emerge which will mirror the peace and justice of God in its original intention.

*   *   *

We have concluded our reflections on the Christian understanding of creation in an age of science and have seen that theology no longer retreats from the encounter with the sciences. It also does not assert its own position in opposition to the findings of the sciences. Since Christian faith is lived in this world and in our present history, the findings of science can be used to illustrate the Christian faith in God the creator, sustainer, and redeemer. In order to do justice to science, this cannot be done by usurping scientific findings for theological purposes, but must take place in continuous dialogue with scientists and their findings.

While science abstracts from the total context of nature to obtain more manageable units which can be investigated under laboratory conditions, human life of any person, whether scientist or theologian, does not exist in these ideal and restricted laboratory conditions. Therefore the questions about origin and destiny, and whether anybody cares, are not just questions peculiar to theological inquiry. They are questions that emerge at one point or other in any of our lives and belong to our very existence as human beings. The question about creation is not just a theoretical one, it is also deeply existential. The assurance and confidence that we are not here by accident, that we know where we are going and that somebody cares, alleviates human anxiety and makes life both bearable and rewarding.

Index of Names

Index of Subjects

Index of Scripture References

254